Experiments and Activities Manual *for*

Principles *of* Electronic Communication Systems

Fourth Edition

Louis E. Frenzel Jr.

Mc Graw Hill Education

EXPERIMENTS AND ACTIVITIES MANUAL FOR PRINCIPLES OF ELECTRONIC COMMUNICATION SYSTEMS, FOURTH EDITION

Published by McGraw-Hill Education, 2 Penn Plaza, New York, NY 10121. Copyright © 2016 by McGraw-Hill Education. All rights reserved. Printed in the United States of America. Previous editions © 2008, 2001, and 1995. No part of this publication may be reproduced or distributed in any form or by any means, or stored in a database or retrieval system, without the prior written consent of McGraw-Hill Education, including, but not limited to, in any network or other electronic storage or transmission, or broadcast for distance learning.

Some ancillaries, including electronic and print components, may not be available to customers outside the United States.

This book is printed on acid-free paper.

2 3 4 5 6 7 QVS/QVS 21 20 19 18 17

ISBN 978-1-259-16648-8
MHID 1-259-16648-1

Senior Vice President, Products & Markets: *Kurt L. Strand*
Vice President, General Manager, Products & Markets: *Marty Lange*
Vice President, Content Design & Delivery: *Kimberly Meriwether David*
Managing Director: *Thomas Timp*
Global Publisher: *Raghu Srinivasan*
Director, Product Development: *Rose Koos*
Product Developer: *Vincent Bradshaw*
Marketing Manager: *Nick McFadden*
Director, Content Design & Delivery: *Terri Schiesl*
Program Manager: *Faye Herrig*
Content Project Managers: *Kelly Hart, Tammy Juran, Sandy Schnee*
Buyer: *Laura M. Fuller*
Design: *Studio Montage, St. Louis, Mo.*
Content Licensing Specialist: *DeAnna Dausener*
Cover Image: *© Royalty Free/Corbis*
Compositor: *Aptara®, Inc.*
Typeface: *10/12 Times Roman*
Printer: *Quad Graphics Versailles*

All credits appearing on page or at the end of the book are considered to be an extension of the copyright page.

The Internet addresses listed in the text were accurate at the time of publication. The inclusion of a website does not indicate an endorsement by the authors or McGraw-Hill Education, and McGraw-Hill Education does not guarantee the accuracy of the information presented at these sites.

Contents

Introduction

This manual is to accompany the textbook *Principles of Electronic Communications Systems* (PECS), fourth edition. That text was written to support a one- or two-semester course in the fundamentals of communications systems including wireless and networking, both analog and digital. The text is written at the technology level and provides coverage for both 2-year community college and technical school programs through 4-year bachelor of technology programs. This manual serves as a supplement to the text and provides practical student activities as well as laboratory experiments, given the appropriate equipment.

There are always limits as to the extent of coverage of any textbook. Given practical page limits in addition to the time constraints of any college course, a text can cover only so much. The scope and breadth of coverage has been balanced with depth of coverage. However, this manual serves the instructor and the student in providing not only validation of principles in the text but also the opportunity to dig deeper on a variety of subjects with both Internet access and hands-on lab experiments.

Organization of this Manual

This manual is divided into two types of projects, activities and lab experiments. The projects are targeted at specific chapters and are numbered as such. For example, Project 3-2 is the second project related to Chapter 3.

Activities

The activities are primarily Internet searches or Web page accesses to acquire additional information on specific topics. Some activities provide questions to answer from the material found. With Web access and search being the dominant method of information gathering on the job in the real-world today, the activities provide students the chance to practice the process they will use considerably in their future employment. Most projects have specific questions to answer by referencing the accessed materials.

The projects typically give relevant websites to avoid excessive search time. Use these websites first to answer the questions. Then use a search via Google, Yahoo, or Bing, using specific search terms to find other answers not in the referenced material.

Lab Experiments

Lab experiments are projects using actual electronic and/or computer hardware to demonstrate and validate key theoretical concepts. The goal is to provide a minimum of true hands-on practice with real test equipment, components, and related hardware.

How to Use this Manual

The projects assume that you have read the related chapter and/or have had one or more lectures on the subject. Instructors will assign relevant projects best fitting their goals for the course. Otherwise, students should feel free to select projects of interest and pursue them.

Lab Equipment

Most colleges have adequate lab equipment for teaching basic electronics as well as digital circuits and microcomputers. However, few have suitable lab equipment for teaching communications hardware. It is typically more expensive than basic instruments and trainers and can certainly be more difficult to use. Working with RF circuits and equipment is particularly challenging.

For these reasons, the lab experiments in the manual have been simplified as much as possible without compromising the teaching of the fundamentals. Furthermore, low cost was a key consideration in choosing the lab experiments. One way to do this is to use circuits that operate at much lower frequencies. Instead of VHF, UHF, and microwave signals that are the most common today, this manual uses low RF and in many cases audio and midrange frequencies below 1 MHz. This makes the circuits less expensive, less critical, and more adaptable to traditional breadboards. An effort has been made to use the older DIP ICs rather than the more common surface-mount devices so common in wireless and communications gear today. Finally, by using the lower frequencies, commonly available test equipment is typically acceptable.

Breadboards and Trainers

Standard breadboards and trainers may be used for most of the experiments described in the manual. The key to making the circuits work is to keep the component leads and interconnecting wires short and the components grouped closer together to minimize the stray inductance, capacitance, and coupling. All power supplies should be bypassed with 0.1-μF ceramic capacitors where they attach to the breadboard to minimize noise, coupling, and other problems.

A good choice for a general-purpose trainer is the National Instruments (NI) ELVIS. See Figure 1. It contains the usual breadboarding sockets as well as multiple power supplies and a function generator. And it is designed to connect to a PC or laptop running NI LabVIEW software that converts the computer into a group of virtual instruments (VIs), which include a multimeter and oscilloscope. This combination is hard to beat since the complete lab station, breadboard, trainer, power supplies, generators, and instruments are contained in two units, the ELVIS trainer itself and the PC. All of the experiments have been tested on the ELVIS trainer. I highly recommend it, as the availability of the LabVIEW software provides useful virtual instruments and allows the demonstration of DSP and other concepts that are impossible with common hardware.

Virtual Instrumentation Trainers

Some colleges are now using virtual test instruments instead of the traditional bench test instruments. A virtual instrument is one that is implemented on a computer. Special circuitry and software is used to create a DMM, oscilloscope, function generator, spectrum analyzer, and even power supplies by using a desktop or laptop computer. The computer screen serves as the readout for the meter, scope, and any other instruments available. On-screen knobs, switches and other controls are manipulated with the mouse.

The ELVIS trainer uses virtual instruments. Two other less expensive virtual instrument products are the National Instruments myDAQ and the Digilent Discovery. See Figure 2. The myDAQ connects to the computer via a standard USB port and derives power to supply DC voltages of 5 V and ±15 V to experimental circuits. In addition, the software referred to as LabVIEW implements a DMM, scope, function generator, and a few other instruments. A connector on the side of the myDAQ provides connections to the power supplies, the instruments, and an external breadboarding socket.

Another virtual instrument is the Digilent Discovery. It too has built-in power supplies, meter, scope, and function generator. The device also uses a USB connection to the computer. Software called *Waveforms* is used on the computer to implement the instruments. Wire leads connect to the breadboarding socket where the experimental circuit is built.

A positive feature of these products is that they are very affordable and typically cost less than a textbook. Many students can afford them. If a student owns a laptop, he or she may want to invest in one of these devices as then they will be able to run many of the experiments in this manual at home, in the dorm, or anywhere when it is convenient. These trainers work well in the lab, and they are affordable if the lab already has computers.

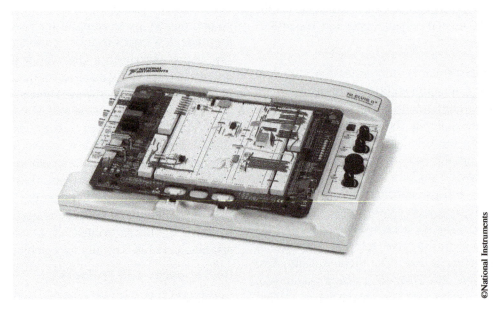

©National Instruments

Fig. 1 The ELVIS trainer. Courtesy National Instruments.

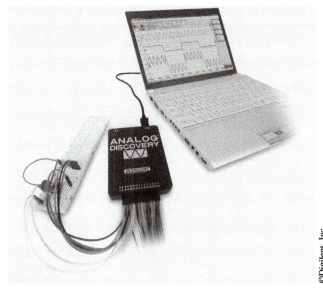

Fig. 2 The National Instruments myDAQ (left) and Digilent Analog Discovery (right) trainers use virtual instruments on a laptop to provide a low-cost base for most lab experiments.

Meters

Any standard digital multimeter is adequate for making tests and measurements in these experiments. It should be capable of measuring dc and ac voltage and current and resistance. A good analog meter is also suitable.

Oscilloscopes

Almost any oscilloscope used in college labs is suitable for use with the experiments in this manual. A dual trace/channel scope is recommended. The bandwidth should be as broad as affordable. I used both 25-MHz and 100-MHz scopes in creating these experiments. However, the virtual oscilloscope in the NI ELVIS trainer, despite its upper bandwidth limit restricted by the sampling rate of the data acquisition card used, is certainly adequate for some low-frequency experiments. Good scope probes are essential for proper results.

Signal Generators

Most of the experiments require one or more signal sources. Since most schools have general purpose function generators that can produce sine and square waves up to about 2 MHz, these are specified in most cases. Some experiments have built-in signal sources to minimize the need for excessive equipment. The ELVIS trainer contains a function generator capable of generating sine, square, and triangular waves up to 250 kHz.

An RF signal generator is highly recommended for some experiments. In general, any good RF signal generator will work, but most of them are very expensive. I strongly recommend the MFJ Enterprises MFJ-269. This instrument is designed for antenna and transmission line tests and measurements. However,

it contains a versatile signal generator that covers the range from 1.8 MHz to 170 MHz and 415–470 MHz. It also has a built-in frequency counter and circuits for measuring SWR. It can measure, calculate, and display complex impedances. Its low cost makes it a good investment for any communications course and lab.

Frequency Counters

Any standard frequency counter will work fine with the experiments described in this manual. The higher the frequency capability, the better. A counter with an upper frequency limit of 200 MHz or more is recommended.

RF Equipment

Two common pieces of RF equipment are recommended, an RF power meter and an SWR meter. In some cases, these units are combined. A recommended low-cost unit is the MFJ-816. The older Bird 43 watt-meters also work very well and are still available. Newer digital watt/SWR meters work well but are not generally worth the expenditure for the few experiments given here.

Other Equipment

One of the most often used instruments in RF and wireless work is the spectrum analyzer. It is, in fact, used more than the oscilloscope in the real wireless world. However, few if any colleges own one. They are hugely expensive. They can be rented, and lower cost used units are available for sale. Even so, the price is high even for a unit with a limited bandwidth.

The experiments in this manual have been designed not to require a spectrum analyzer, but it is highly educational when it can be used to view harmonics, sidebands, and the effects of modulation and noise.

I highly recommend that the college invest in at least one spectrum analyzer and use it in as many of the experiments as possible to give the students the experience that will serve them well later on the job.

Components

The components required are listed at the beginning of each experiment. Only commonly available, low-cost components are specified. They are available locally from distributors in the larger cities or from mail order sources like Digi-Key and Jameco Electronics. These latter two sources are highly recommended for their low prices and rapid delivery.

Some of the experiments are design projects that will require values that cannot be previously determined. As a result, it is good lab practice to buy and keep on hand complete kits of standard-value resistors and capacitors. One-quarter-watt resistors are satisfactory for most designs. Ceramic and polystyrene capacitors are recommended. Both the mail order suppliers mentioned above stock standard kits that are very affordable and very handy in any lab.

Kits

A few of the projects require the purchase of kits, assemblies of parts designed to build a specific piece of equipment. Kits provide a low-cost way to implement the more complex experiments without problems. They also provide students with experience in building actual circuits on printed circuit boards, using real ICs and other components.

The kits recommended for these projects are very low in cost and many can be purchased for a given class. The students may even want to purchase them for themselves. In any case, building the kits can be part of the lab experience. Later, the assembled kits may be retained and used in future classes without the need to build them.

Commercial Lab Equipment

Another good choice for a communications laboratory is to use commercial training equipment. Listed below is a sampling of companies and some of their communications lab products worth considering.

- EMONA tims (www.emona-time.com)
  - Stand-Alone Telecoms Trainer
  - Telecoms Trainer (uses NI ELVIS)
  - Fiber Optics Trainer (uses NI ELVIS)
  - Signal-Processing Trainer (uses NI ELVIS)
- Feedback Instruments (www.feedback-instruments.com)
  - Mobile Phone Trainer
  - Bluetooth Trainer
  - ZigBee Trainer

- USB Trainer
- RFID Trainer
- Embedded Internet Trainer
- CAN Bus Trainer
- Analog and Digital Telecommunications Workstation
- Amplifiers and Oscillators Workboard
- Tuned Circuits and Filters Workboard
- Modulation and Coding Workboard
- Microwave Trainer
- MIDE Microwave Design Software
- Antenna Lab
- Complete Microstrip Trainer
- Antenna System Demonstrator
- Transmission Line Demonstrator
- LAN-T Trainer
- Fiber Optics Trainer
- Optical Fiber Trainer
- Optical Fiber System
- Keysight Technologies (formerly Agilent Technologies, www.keysight.com)
  - TDR Probe Kit
  - Lightwave Laboratory Kit
  - RF Laboratory Kit
  - RF Transceiver and Courseware
  - Series 1000 Oscilloscope Training Kit
  - Analog Electronics Lab Station
  - Digital Signal Processing Lab Solution
  - Digital RF Communications Training Kit
- Lab-Volt Systems Inc. (www.labvolt.com)
  - Radar Training Systems
  - Satellite Communications Training Systems
  - Antenna Training and Measuring System
  - Telephony Training Systems
  - Microwave Technology Training System
  - Communications Technologies Training Systems
  - Digital Communications Training Systems
  - Analog Communications Training Systems
  - Microwave Technology Training System
- National Instruments (www.ni.com)
  - ELVIS Computer-Based Instrumentation
  - Universal Software Radio Peripheral (USRP) Platform
- Nida Corporation (www.nida.com)
  - 360S Systems Trainer
  - Training Console
  - Basic Microwave
  - Microwave Communications
  - Microwave Standing Wave Ratio Measurement
  - Microwave Reflection
  - Radio Trainer

Chapter

1

Introduction to Electronic Communication

Project 1-1 Exploring the Regulatory Agencies

Objective

Become familiar with those U. S. government agencies responsible for managing and regulating electronic communications.

Introduction

Both the Federal Communications (FCC) and the National Telecommunications and Information Administration (NTIA) are those government agencies designated to control and regulate electronic communications and manage the wireless spectrum in the United States. Everyone working in the communications field must absolutely be familiar with these organizations, their purpose, and procedures. This project provides an introduction to both.

Required

PC with Internet access.

Procedure

1. Go to the website of the FCC at www.fcc.gov.
2. Explore the website to see what it contains.
3. Repeat the procedure by going to the NTIA website at www.ntia.doc.gov.
4. Explore the website to see what it contains.
5. Using both websites as your information sources, answer the questions below. Use the search feature available on each site to find things you are interested in.

Questions

1. To whom does the FCC report within the government organization?
2. To whom does the NTIA report within the government organization?
3. Describe the purpose and function of the FCC.
4. Describe the purpose and function of the NTIA.
5. What is the FCC CFR Title 47?
6. Search the FCC website for information on licensing. Locate the information on commercial operators' licenses. What is the GROL? Who is required to have one?
7. What is the procedure for getting a GROL?

Project 1-2 Becoming Familiar with the FCC Rules and Regulations

Objective

Access the FCC rules and regulations pertaining to wireless and other forms of electronic communications and learn how to find specific answers to regulatory questions.

Introduction

The FCC rules and regulations are extensive and anyone working in communications needs to be familiar with them. It is easy to violate these rules, and that can lead to major troubles for you personally or your company or its vendors, clients, or customers. Fines and even jail time can occur for really serious violations. Lack of knowledge of these rules and regulations is no excuse for their violation.

This project will let you see just how large and complex the regulations are. The goal is mainly to provide you with the knowledge and skills of accessing the rules and applying them to your work.

Required

PC with Internet access.

Procedure

1. Go to the FCC website www.fcc.gov and click on the rules and regulations link on the left side of the page.
2. From the next page, access the FCC rules Title 47 of the Code of Federal Regulations (CFR). This will bring you to a list of the different parts of the rules that pertain to specific sectors of communications. Scan through this list and become familiar with the different categories.
3. Using the rules as your source of information, answer the questions below.

Questions

1. What is contained in Subpart A of Part 2? Will you find that useful?
2. In Part 2, state what communications services can occupy the spectrum from 108 to 138 MHz?
3. What part pertains to commercial radio operators and licensing?
4. Give a brief overview of the kinds of devices covered by the Part 15 regulations.
5. Give a brief overview of the kinds of devices covered by the Part 18 regulations.
6. What part covers TV broadcasting?
7. Is cable TV regulated by the FCC? Which part?
8. Under what part(s) is radar covered?
9. Which parts cover CB and FRS personal radio services. What is the operating frequency ranges and power limits for each service?
10. List all of the communications modes and modulation methods allowed under the amateur radio regulations.
11. Under which part is radio-frequency interference and electromagnetic interference explained?
12. Which parts cover cell phones and wireless local area networks?

2

Project 1-3 Exploring Communications Applications

Objective

Identify unique and interesting uses of wireless and network communications.

Introduction

Chapter 1 listed a wide range of applications for wireless and communications. The categories given are broad and general. Yet there are an amazing number of unusual and special applications not listed. Your assignment here is to identify at least three new or different applications for wireless and wired networking not given in the text. You may do this from general knowledge, asking others, researching magazines, or by any other means. One or more of these should come from doing Internet searches.

Required

PC with Internet access.

Procedure

1. Identify unique and special wireless applications. List at least three and describe briefly, in a sentence or two, the purpose and benefit of each.
2. Identify unique and special wired networking applications. List at least three and describe briefly, in a sentence or two, the purpose and benefit of each.

Project 1-4 Wireless Technology Overview

Objective

Review wireless methods and applications.

Introduction

Communications techniques use both radio and wired media for sending and receiving information. While communications began with wired techniques such as telegraph and telephone, today wireless techniques dominate. This project provides you with a general overview of how wireless works and the more common wireless applications covered later in the text.

Required

PC with Internet access.

Procedure

1. Go to the website www.work-readyelectronics.org.
2. At the top of the page, click on the Modules box, which will take you to a second page listing all the various modules you can access for free.
3. Answer the three questions given, then select the Wireless Technology module.
4. When you have accessed the module, follow the instructions given and complete the module.
5. Be sure to take advantage of the Knowledge Probes that help you to review the material.
6. Take the formal Assessment at the end of the module.

Chapter 2

Review of Electronic Fundamentals

Project 2-1 *LC* Low-Pass Filter Design and Testing

Objective

Design, build, and test a low-pass *LC* filter.

Introduction

There are many different types of filters used in wireless and other communications applications. Newer types such as SAW and DSP are widespread, but the older traditional *LC* filters are still widely used in many designs. They are so widespread that you must have some familiarity with their design, application, and testing. Providing design details is beyond the scope of this book, but there are Internet resources that can perform this task. In this project, you examine several filter design websites and resources, then use one to design a filter. You will then build that filter with real components and test it. Finally, you will test the filter to become familiar with general specifications and measurement practices.

Required

1 PC with Internet access.
1 breadboard trainer
1 function generator
1 oscilloscope
• Resistors, capacitors, and inductors as specified by the designs

Procedure

1. Do a Web search using Google or Yahoo on the term "filter design." Scan through the recovered sites to see what is available. What two types of filters are the most commonly referred to in these sites?
2. Go to the website www-users.cs.york.ac.uk/fisher/lcfilter/. Explain what you find there and the procedure described.
3. Go to the website www.aade.com/filter32/download.htm. Download the filter design software and install it on the PC you are using.
4. Use the AADE software as described on the website to design a low-pass *LC* filter with the following specifications.
   a. Cutoff frequency of 32 kHz.
   b. 1-kΩ input and output impedances.
   c. Pi-type Butterworth design.
   d. Three poles or an attenuation of at least 30 dB at 100 kHz.
5. Design the filter and print out the schematic and parts list.

6. Use the analysis feature to test the filter and print out the frequency plot.

7. Build the filter, using the closest standard component values. In most cases, the design will produce nonstandard values of capacitance and inductance. In some cases, you can get closer to the design values by combining two or more standard-value capacitors in parallel.

8. Test the filter by running a frequency-response curve, using the function generator and oscilloscope. Plot your frequency curve on standard logarithmic graph paper. You can also use the Bode plot feature in ELVIS if you are using NI LabVIEW.

9. What is the passband insertion loss of this filter at 30 kHz? Find that value on the breadboarded circuit.

10. What is the roll-off rate of this filter?

Project 2-2 *RC* Active Filter Design and Testing

Objective
Design, build, and test an *RC* active band-pass filter.

Introduction
For audio and below 1-MHz applications, *LC* filters are no longer widely used because of the high cost, size, and weight of inductors for these low frequencies. At these frequencies, *RC* active filters based on op amps are popular and commonly used. They are relatively easy to design and build. In this experiment, you will use a common technique to design a typical *RC* active filter and then breadboard it and test it.

Required
1 breadboard trainer
1 dual $\pm$12-V power supply
1 multimeter
1 oscilloscope
1 function generator
1 op amp IC (any common device will work such as the 741, 1458, TL072, or 301)
• Resistors and capacitors as determined by the design

Procedure
1. Go to the website www.ti.com. Locate Application Report SLOA093 (December 2001). Its title is *Filter Design in Thirty Seconds* by Bruce Carter. Print out a copy of this paper and read it.
2. Use the procedure described to design a narrow bandpass filter for 19 kHz. Use the procedure for a dual ($\pm$) supply op amp. Acquire the parts.
3. Breadboard the circuit.
4. Using a function generator and oscilloscope, plot the frequency-response curve of the filter. You can also use the Bode plot feature in LabVIEW if you are using it and the ELVIS trainer.
5. Make any necessary measurements to determine the following:
   a. Filter insertion loss (or gain) at the center frequency.
   b. Filter bandwidth.
   c. Calculate filter *Q*.
   d. What is the maximum value of input voltage that can be applied before clipping (distortion) occurs?

Project 2-3 Ceramic Band-Pass Filter

Objective

To demonstrate the operation and selectivity of a typical ceramic filter and to show a method of plotting frequency-response curves.

Required

1 oscilloscope
1 function generator capable of sweep frequency operation
1 frequency counter
1 455-kHz ceramic filter with 8-kHz bandwidth (Toko Type CMF2 or equivalent)
2 2-kΩ resistors
2 10-kΩ resistors
1 1N34 or 1N60 germanium diode
1 0.1-μF ceramic capacitor

Introduction

Most communications receivers, regardless of size or application, are super-heterodyne types. In addition, most of the gain and virtually all of the selectivity of the receiver is obtained in the IF section. The gain is easily obtained with transistor or IC amplifiers. The selectivity is more difficult as the IF passband must be carefully selected to match the signals being received and must, in most cases, have very steep skirt selectivity. In older receivers, selectivity was obtained with cascaded tuned transformers whose coupling was adjusted to give the desired bandwidth and steepness of response. These IF transformers were difficult to tune and adjust.

Today, most IF selectivity is obtained with crystal filters or ceramic resonators. Some designs even use DSP filters. At very high frequencies, SAW filters are used. These filters are fixed during their manufacture to furnish a desired frequency response. Very steep selectivity is readily obtained in a very small package at low cost, and no tuning or adjustment is required.

In this experiment, you will demonstrate the operation of a typical IF ceramic filter. You will show its response by tuning and show its attenuation of the signal. Then you will observe a method that lets you plot the frequency-response curve directly on the oscilloscope.

Procedure

1. Build the circuit shown in Fig. 2-1.
2. Connect an oscilloscope across the output. Turn on the function generator and set its frequency to approximately 455 kHz. Adjust the input voltage to the filter to 4 V_{P-P}.
3. Tune the function generator frequency until you notice a peak in the output voltage across the load resistor. Tune the frequency slowly and carefully until maximum output voltage is obtained. Measure the output voltage and record the value below. Measure the frequency at which the peak output voltage is obtained. Be sure to use a frequency counter for best accuracy.

   $f =$ _____
   $V_0 =$ _____
   Does the filter have more than one peak?
4. Calculate the gain or loss of the filter in dB.

   dB = _____
5. Slowly tune the function generator frequency above and below the filter frequency of 455 kHz. Note how fast or how slow the output voltage changes. Describe in your own words how the output changes with frequency.
6. Refer to Fig. 2-2. Add the diode and its 10-kΩ load and the low-pass filter made up of the 10-kΩ resistor and the 0.1-μF capacitor.

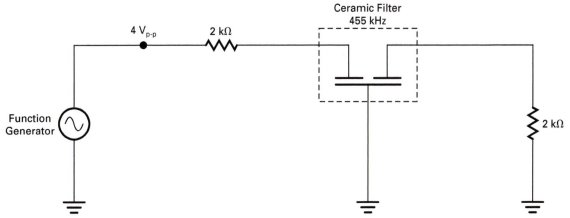

Fig. 2-1 Filter measurement circuit.

7. Set the function generator frequency to a value of about 400 kHz. Then, set up the generator for sweep frequency operation. Refer to the generator operation manual or have your instructor explain the sweep mode. Most generators frequency modulate their internal oscillator with a linear sawtooth wave, which varies the output frequency from some lower value continuously to some higher value. You will apply this linearly increasing frequency to the filter. You will observe the output of the diode detector at the filter output.

8. While observing the output of the diode detector, adjust the sweep characteristics and the oscilloscope time base until you display a waveform that is the frequency-response curve of the filter. Draw what you see. What is it?

Questions

1. Explain how the diode detector and swept frequency combine to plot the frequency-response curve of the filter.

2. How could you determine the 6-dB bandwidth of this filter? Describe the process and then implement the procedure on the experimental circuit and state the bandwidth you measure. Be sure to use a frequency counter for your measurements.

3. Describe the shape of the response curve in the pass band. Is it a single peak, a flat horizontal line, or does it have some variation?

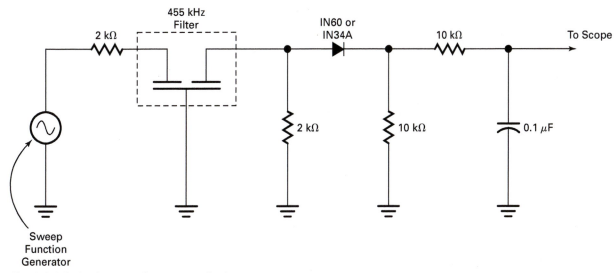

Fig. 2-2 Diode detector for sweep display.

Project 2-4 Switched Capacitor Filters

Objective

To demonstrate IC switched capacitor filters.

Required

1	oscilloscope (dual-trace, 20-MHz minimum bandwidth)
3	function generators
1	frequency counter
1	dc power supply capable of supplying ± 5 V
1	general-purpose op amp IC (741, 411, etc.)
3	100-kΩ resistors
1	MF10 programmable switched capacitor filter ICs (National Semiconductor, now Texas Instruments, or Maxim)
1	74LS93 TTL binary counter IC
1	22-kΩ resistor
1	33-kΩ resistor
2	220-kΩ resistors
2	330-kΩ resistors
2	0.1-μF capacitors

References

1. *Principles of Electronic Communication Systems,* Chapter 2.
2. MF10 data sheet (National Semiconductor, now Texas Instruments, or Maxim).

Introduction

An alternative to conventional *RC* or *LC* passive filter or *RC* active filters is switched capacitor filters (SCFs). This IC filter uses a capacitor network switched by fast MOSFETs. When combined with op amps in a feedback arrangement, the result is a very selective filter. Most SCFs can be programmed to function as a low-pass, high-pass, band-pass, or notch filter. This all-electronic filter is smaller and lighter than conventional LC filters, and it can be programmed for a specific frequency and bandwidth.

SCFs are sampling filters in that they take in an analog signal and sample it at a rate many times the highest frequency content. Samples of the signal voltage are stored in the capacitor network. A clock oscillator sets the sampling rate. The frequency of the clock sets the center or cutoff frequency of the filter.

The characteristics of a SCF are fully adjustable by setting the clock frequency. The Q and therefore the bandwidth (BW) of the filter are also programmable. A variety of low-cost SCFs are available in IC form. They can be used to form low-pass, high-pass, and band-pass and notch filters in the range 0 to 1 MHz. You will demonstrate the popular MF10 SCF in this experiment.

The MF10 consists of two filters in a 20-pin-DIP IC. Externally connected resistors select the filter mode and operational characteristics. The design procedures are beyond the scope of this experiment, but more details can be obtained from the manufacturer's data sheet. In this experiment, the MF10 is programmed as a band-pass filter (BPF). Its Q is determined by the external resistor ratios. As in any BPF, the bandwidth (BW) is

$$BW = \frac{f_{CLK}}{Q}$$

The filter center frequency is f_0.

The center frequency of the BPF is determined by the clock frequency (f_{CLK}) applied to the internal MOSFET sampling switches:

$$f_0 = \frac{f_{CLK}}{50}$$

For example, if the clock is set to 30 kHz, the BPF center frequency is 30 kHz/50 = 600 Hz. The clock signal is a TTL-compatible 50 percent duty cycle square wave. Its maximum value is 1 MHz. This sets the upper frequency limit of the center frequency to

$$f_0 = \frac{1000\ kHz}{50} = 20\ kHz$$

In this experiment, you will demonstrate the filter operation and characteristics.

Procedure

SCF Operation

1. Construct the MF10 filter circuit shown in Fig. 2-3. Remember that there are two filters in the IC and you will test both of them. The input to the first filter is labeled V_{inA} at the 220-kΩ resistor R_1. The output of the first filter is at pin 2 and is labeled V_{oA}. The input to the second filter is labeled V_{inB} at the 330-kΩ resistor. The output of the second filter is at pin 19

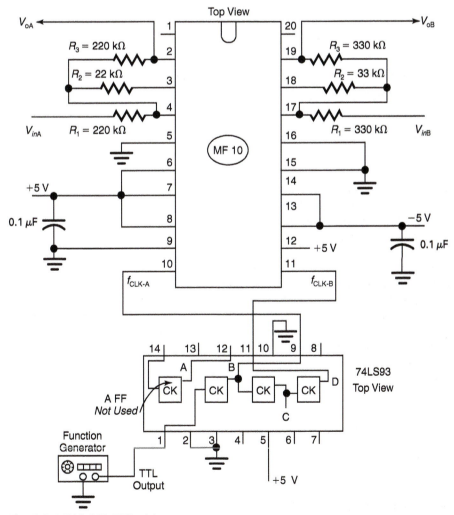

Fig. 2-3 MF10 SCF BPF wiring.

and is labeled V_{oB}. The gain of the filter is the ratio of $-R_3/R_1$. The Q of the filter is the ratio of R_3/R_2.

The 74LS93 is a TTL binary counter used as a frequency divider. The clock signal is derived from an external function generator that drives the counter. The output of the first flip-flop in the counter (pin 9) divides the clock signal by 2 to ensure a 50 percent duty cycle signal to the f_{CLK-A} input that sets the center frequency of the first filter. The output of pin 11 on the 74LS93 divides the f_{CLK-A} by 4 to get the clock signal for the second filter, f_{CLK-B}. A tunable function generator with a TTL output is connected to pin 1 of the 74LS93 to provide the clock signal. Set the clock generator frequency to 500 kHz. Use the frequency counter to ensure setting accuracy.

2. What is the center frequency of the filter?

$f_0 = $ _____

3. Connect a function generator to the V_{inA}. Set the generator to supply a 2-V_{p-p} sine wave to the filter. Monitor the filter output at pin 2 on the oscilloscope. Monitor the function generator output on the frequency counter.

4. Tune the input sine-wave function generator frequency until you see the sine-wave voltage at the filter output peak. Measure the frequency and the output voltage.

$f_0 = $ _____ $\qquad$ $V_{oA} = $ _____

Does the center frequency of the BPF correspond to the value you calculated in step 2?

5. Look at the filter output signal closely. Describe the signal. How is it different from a true sine wave?

6. Using the peak output voltage in the step above, determine the 23 dB down value. Voltage at -3 dB down = _____

7. Tune the input signal below the center frequency until the output voltage drops to the -3 dB value. This is the lower cutoff frequency (f_L). Measure this value on the counter and record below. Then tune the input signal above the center frequency until you again see the -3 dB down value. Note the frequency, as this is the upper cutoff frequency (f_U).

$f_L = $ _____ $\qquad$ $f_U = $ _____

8. Calculate the bandwidth of the filter.

BW = _____

9. Now, calculate the Q of the filter.

$Q = $ _____

10. Repeat steps 2 through 8 for the second filter. Set the external clock frequency to 160 kHz. Use the counter to measure it for accuracy. Then connect the sine-wave function generator to the V_{inB} on the MF10 and observe the filter output at pin 19.

$f_0 = $ _____

$f_0 = $ _____

$V_{oA} = $ _____

Voltage at -3 dB down = _____

$f_L = $ _____

$f_U = $ _____

BW = _____

$Q = $ _____

Questions

1. Assume a MF10 filter with $R_1 = 47$ kΩ, $R_2 = 22$ kΩ, $R_3 = 47$ kΩ. The clock frequency is 72 kHz. Find the filter center frequency, Q, and the bandwidth.

2. What would be the effect of cascading the two filters in the MF10?

3. How is the center frequency of the filter varied?

4. (True or False) The MF10 can be used for low- and high-pass filtering.

Project 2-5 Fourier Theory Review

Objective

Review Fourier theory.

Introduction

Fourier theory is one of those mathematical processes that is essential to the study and understanding of any communications method. While a strong math background is desirable for learning the more advanced concepts of this broad mathematical theory, you can learn the Fourier concepts with just a good algebra and trigonometry background. Because it explains so much of what occurs in digital systems and wireless, it is important that you understand this key mathematical concept. This project provides a review of the material covered in the text.

Required

PC with Internet access.

Procedure

1. Go to the website www.work-readyelectronics.org.
2. At the top of the page, click on the Modules box, which will take you to a second page listing all the various modules you can access for free.
3. Answer the three questions given, then select the Fourier module.
4. When you have accessed the module, follow the instructions given and complete the module.
5. Be sure to take advantage of the Knowledge Probes that help you to review the material.
6. Take the formal Assessment at the end of the module.

Chapter

3

Amplitude Modulation Fundamentals

Project 3-1 Measuring the Percent of Modulation

Objective

To learn two methods of expressing and measuring the percent of modulation of an AM signal and to observe the condition of overmodulation.

Required

1   oscilloscope with provision for external horizontal input
1   frequency counter
1   12-V dc power supply
1   function generator with sine output
1   XR-2206 Exar function generator IC
1   0.002-μF capacitor
1   0.47-μF capacitor
1   1-μF capacitor
1   10-μF electrolytic or tantalum capacitor
2   4.7-kΩ resistors
1   6.8-kΩ resistor
1   10-kΩ resistor
1   47-kΩ resistor
1   150-kΩ resistor
1   1-kΩ pot
1   10-kΩ pot

References

Principles of Electronic Communications Systems, 3rd ed., Chapter 3.

Introduction

Optimum amplitude modulation occurs when the peak-to-peak value of the modulating signal is equal to the peak value of the carrier. In that way, the modulating signal causes the carrier amplitude to vary from twice its unmodulated value to zero, as Fig. 3-1 illustrates. That is called a modulation index m of 1, or 100 percent. Of course, a modulation index of 0, or 0 percent modulation, occurs when the modulating signal amplitude is 0 and a constant-amplitude carrier signal results. Intermediate values of modulation index or percent can be determined by measuring the amplitudes of the

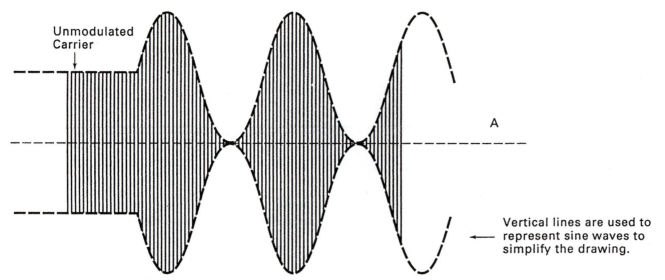

Fig. 3-1 100 percent modulation.

Vertical lines are used to represent sine waves to simplify the drawing.

modulated signal, as shown in Fig. 3-2. The following formulas can then be used to compute the percent of modulation:

$$\text{Modulation index } m = \frac{V_{max} - V_{min}}{V_{max} + V_{min}}$$

$$\text{Percent modulation} = m \times 100$$

The most important criterion is to attempt to achieve 100 percent modulation but not exceed it. If the modulating-signal amplitude is greater than the carrier amplitude, overmodulation will occur and the information signal will be distorted. Note how the peaks in Fig. 3-3 are clipped.

Another way to measure the modulation index or percent of modulation is to use a trapezoidal pattern that can be displayed on an oscilloscope. Such a pattern is shown in Fig. 3-4. To create it, the AM signal is connected to the vertical input as usual. The modulating information signal is connected to the external horizontal input and is used as the sweep signal. The scope is put into the external sweep or x vs. y display mode, and the vertical and horizontal gains are adjusted until the pattern appears. The V_{max} and V_{min} values can then be read from the pattern and used in the previously given formula to calculate the modulation index and percent of modulation.

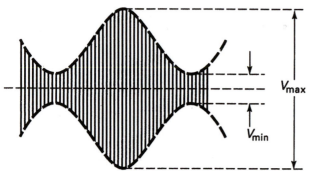

Fig. 3-2 Measuring voltages to calculate modulation index.

16

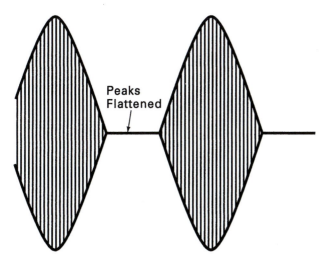

Fig. 3-3 Overmodulation-caused distortion.

The trapezoidal pattern method is usually preferred because measurement of V_{max} and V_{min} is easier. Further, the slanted sides of the trapezoid readily show distortion. Straight sides mean no distortion; nonlinear sides indicate overmodulation or other distortion.

To demonstrate AM, you will use the XR-2206 Exar function generator IC. This chip contains an oscillator that produces sine, square, triangle, and sawtooth waves. The frequency is set by an external resistor and capacitor. The oscillator can also be voltage-controlled, which permits its frequency to be varied. This allows sweep frequency or frequency modulation to be produced, but that function will not be used here.

The chip also contains an amplitude modulator circuit. The internal oscillator serves as the carrier, which can be amplitude-modulated by an external input signal. The modulator produces excellent AM, but the modulating signal amplitude required is higher than the theoretical value necessary to produce a given degree of modulation.

Procedure

1. Construct the circuit shown in Fig. 3-5. You will use the external signal function generator with a 1-kHz output to modulate the carrier produced by the 2206 IC function generator. The 2206 chip contains a built-in amplitude modulator.

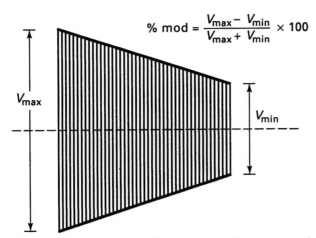

$$\% \text{ mod} = \frac{V_{max} - V_{min}}{V_{max} + V_{min}} \times 100$$

Fig. 3-4 Trapezoid pattern for measuring the percent of modulation.

17

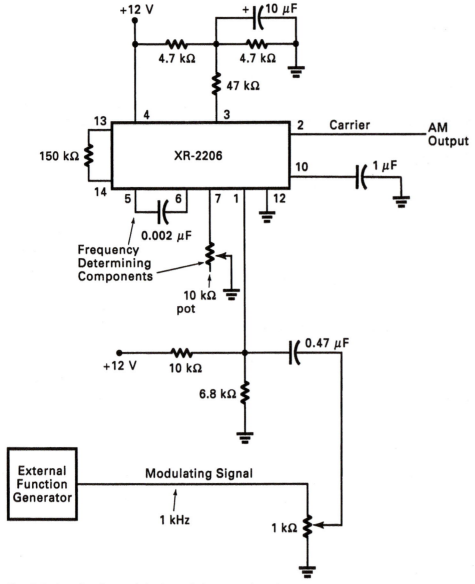

Fig. 3-5 Amplitude modulation of the 2206 function generator IC.

2. Apply power to the circuit. Adjust the dc supply voltage to 12 V. Observe the waveform at the output at pin 2 of the carrier function generator IC. Reduce the modulating signal input from the external function generator to zero with the 1-kΩ pot or the amplitude control on the function generator. While measuring the output at pin 2 with a frequency counter, adjust the 100-kΩ pot for a frequency of 40 kHz. (*Note:* If no counter is available, use the calibrated horizontal sweep of the scope to set the frequency.)

3. Measure and record the carrier output voltage, peak and peak-to-peak, with the oscilloscope.

4. Use the value obtained in step 3 to calculate and record the amount of voltage required to 100 percent modulate this carrier.

5. Vary the amplitude of the modulating signal from the external function generator by using the 1-kΩ pot and/or the external amplitude control from zero to about 5 V$_{p\text{-}p}$. Note the effect on the AM signal.

6. Adjust the modulating signal amplitude for maximum value just prior to the occurrence of distortion. If clipping like that shown in Fig. 3-3 occurs, reduce the input signal amplitude until the distortion just disappears. Measure and record V_{max} and V_{min}.

7. Calculate and record the modulation index and percent of modulation. Can 100 percent modulation be achieved without clipping?
8. Measure and record the amplitude, both peak and peak-to-peak, of the modulating signal voltage. How does this value compare to the value you calculated in step 4?
9. Display the trapezoidal pattern to measure percent modulation. Leave the vertical scope input on the AM signal at pin 2. Connect the modulating signal at the 1-kΩ pot output to the external horizontal (sweep) input. Set the time base control or other switch on the scope for external input sweep. Adjust the scope vertical and horizontal gain controls to display the trapezoid.
10. Vary the modulating signal amplitude and note the variation in the trapezoid. Vary the horizontal and vertical gain controls on the scope to keep the pattern centered.
11. Temporarily set the scope back to normal time base operation. Observe the AM wave. Adjust the modulating signal amplitude for 100 percent modulation without distortion.
12. Return the scope to the previous condition with external horizontal sweep. Observe the pattern. What is the shape of the pattern with 100 percent modulation?
13. Return the scope to normal time base operation. Observe the AM wave.
14. Adjust the amplitude of the modulating signal to zero so that the output is a constant-amplitude sine wave. Measure and record the carrier amplitude, both peak and peak-to-peak.
15. Set the input modulating signal amplitude to 0.5 V_{p-p}.
16. Calculate and record the modulation index and modulation percent you would expect from the voltages given in steps 14 and 15.
17. Now view the AM signal on the oscilloscope and determine V_{max}, V_{min}, and the percent of modulation. Record your observations.
18. Compare the values you obtained in steps 16 and 17. Explain any discrepancies.
19. While observing the AM signal at pin 2 on the 2206 IC, increase the amplitude of the modulating signal to achieve 100 percent modulation. Then continue increasing the modulating signal amplitude until distortion occurs. Describe the effect you observe when overmodulation occurs.
20. Do not disassemble the modulator circuit, because you will use it in future experiments. For now, turn off the power.

Questions

1. On an AM wave, V_{max} = 2 V and V_{min} = 0.5 V. The modulation index is
   _____ .
   a. 0.25
   b. 0.4
   c. 0.6
   d. 0.85
2. For 100 percent modulation, what is the relationship between the carrier voltage V_c and the modulating signal voltage V_m?
   a. $V_m = V_c = 0$
   b. $V_m = V_c$
   c. $V_m > V_c$
   d. $V_m < V_c$
3. When overmodulation occurs, _____ .
   a. $V_m = 2V_c$
   b. $V_m = V_c$
   c. $V_m < V_c$
   d. $V_m > V_c$

4. When you are using a trapezoidal pattern to measure modulation percentage, 100 percent modulation is indicated by what shape of pattern?
   a. Triangular
   b. Rhomboid
   c. Square
   d. Rectangle
5. (True or False) The frequency of the modulating signal affects the percent of the modulating signal.

Project 3-2 Spectrum Analyzer

Objective

To demonstrate the operation and application of a spectrum analyzer.

Required

All the equipment and circuitry used in Project 3-1 plus a spectrum analyzer capable of measuring and displaying low-frequency signals.

Introduction

A spectrum analyzer is a test instrument designed to measure and display signals in a frequency-domain format. Like an oscilloscope, the spectrum analyzer uses a CRT or LCD display, but unlike a scope, it displays signal amplitude with respect to frequency rather than time. Complex waveforms made up of many signals are shown as vertical lines whose heights indicate signal strength and whose horizontal positions indicate frequency. In other words, the spectrum analyzer effectively displays a complex signal as its individual sinusoidal components.

The spectrum analyzer is an ideal test instrument for use with modulated signals because it allows you to see the carrier and sidebands resulting from the modulation. It makes measurement and troubleshooting faster and easier.

In this experiment, you will become familiar with spectrum analyzer operation and use. You will display the spectrum produced by the 2206 amplitude-modulated function generator you analyzed in the preceding experiment.

Procedure

1. Acquire the operation manual for the spectrum analyzer you will be using. Take some time to read and study it. Concentrate on the following:
   a. How does the spectrum analyzer work? Analyze its operation in terms of a block diagram of the major circuits.
   b. How do you use the spectrum analyzer? What are the various controls, inputs, and output? What are the procedures for setting up, calibrating, and making measurements with the unit?
   c. What are the unit's specifications?
2. Set up the spectrum analyzer and become familiar with the controls and how to display a signal.
3. What display would you expect to see on the spectrum analyzer display if a 1-kHz square wave is connected? Sketch the frequency domain display you would see.
4. Connect a 1-kHz square wave from a function generator to the spectrum analyzer. Adjust the controls to see the display. Draw what you see. Is it the same as you predicted in step 3? Explain.
5. Apply power to the 2206 function generator IC you constructed in Lab Experiment 2-1. Adjust the output for a frequency of 35 kHz. Set the modulating signal frequency to 1 kHz. While observing the output at pin 2 of the 2206 (see Fig. 3-5), increase the modulating signal amplitude until 100 percent modulation is achieved. Predict what signal the spectrum analyzer will display and sketch it. (*Note:* If your spectrum analyzer will not accommodate the 35-kHz carrier, reduce the carrier frequency by connecting a larger capacitor to the XR-2206 in place of the 0.002-μF unit on pins 5 and 6.)
6. Apply the AM signal to the spectrum analyzer and adjust the controls to display the signal. How does it compare to the prediction you made?
7. Reduce the modulating signal amplitude and note the effect if has on sideband amplitude.

8. While observing the AM signal, increase the modulating signal amplitude until overmodulation occurs. The signal should be as clipped and distorted as possible. Observe the spectrum analyzer display. Explain how the distortion is indicated.

Questions

1. (True or False) The spectrum analyzer displays the amplitudes of the individual sinusoidal components of a complex signal as determined by the Fourier theory.

2. With 60 percent modulation and a 500-mV carrier, the sideband amplitude for a sine-wave modulating signal is _____ .
   a. 150 mV
   b. 250 mV
   c. 300 mV
   d. 500 mV

3. The largest component in an AM signal as displayed on a spectrum analyzer is the _____ .
   a. Upper sideband
   b. Lower sideband
   c. Carrier
   d. Modulating signal

4. Distortion from overmodulation is displayed by a spectrum analyzer as _____ .
   a. Greater carrier power
   b. Higher sideband amplitudes
   c. Lower carrier and sideband amplitudes
   d. Additional harmonic-related sidebands

5. The horizontal sweep calibration of a spectrum analyzer is given in units of _____ .
   a. Time
   b. Frequency
   c. Voltage
   d. Power

Project 3-3 AM Radio Experience

Objective

To tune and listen to radio stations using amplitude modulation.

Required

1 AM broadcast radio
1 receiver capable of receiving shortwave (SW) signals*
1 receiver capable of receiving citizen band (CB) stations*
1 receiver capable of receiving VHF aircraft radio stations (scanner radio)
1 receiver covering the amateur radio bands at 80, 40, or 20 m and capable of receiving SSB transmissions (built-in BFO and product detector)*
• Antennas as required

*Shortwave (SW) radio.

Introduction

AM is the oldest form of modulation and, despite its disadvantages, it is still used in many communications applications. AM broadcast radio is the most common example. AM is also the most common modulation for shortwave (SW) broadcast radio. Two examples of two-way radio are citizen band (CB) radio and aircraft radio. CB radio is used by individuals for car-to-car communications and by truckers because it is effective and very inexpensive. And it is available to anyone with the money to purchase a transceiver. AM is also still used in aircraft radios. These are the VHF radios used from tower to airplane and from plane to plane. Again, they are low-cost and very effective. A common use of the SSB form of AM radio is amateur radio communications.

The purpose of this experiment is to listen to typical AM transmissions and gain experience using typical communications equipment.

Procedure

1. State the frequency range of the AM broadcast band and the frequency spacings of stations.

   _____

   _____

2. Use an AM radio to listen to the broadcast band. Do this during daylight hours. Tune the entire extent of the band, count the number of stations you hear, and record their frequencies.

   _____

   _____

3. Listen for weak stations. What is the main problem in hearing weak stations?

   _____

   _____

4. Repeat steps 2 and 3 at night. What are the main differences between day and night reception?

   _____

   _____

5. What is the frequency range of the CB band? How many channels are available?

   _____

   _____

6. Use the receiver in a CB transceiver or a scanner radio to monitor the CB channels. Which channels are most widely used?

_____

_____

7. Try out the squelch and noise blanker controls on the receiver.

_____

_____

8. Monitor the channels and try to determine the main use of the CB band. What are some common applications?

_____

_____

9. State the frequency range of the aircraft radio band.

_____

_____

10. Use a scanner radio to monitor these aircraft frequencies. Listen to as many different conversations as you can. Describe briefly what you hear.

_____

_____

11. Try out the squelch and noise blanker controls on the receiver.

_____

_____

12. Use an amateur radio transceiver or receiver to listen to SSB transmissions. A good example is the continuous communications on the amateur 20-m band. Tune the 14.2- to 14.3-MHz range with the receiver BFO on. As you tune the receiver slowly, note how the tuning affects the voice. Explain what you are hearing.

_____

_____

13. If a USB/LSB switch is available on the receiver, try out both positions to see which one produces the most signals.

_____

_____

14. Tune the ham bands for Morse code signals. Use the BFO. Some frequency ranges to try are 3.5–4, 7–7.3, 10.0–10.15, 14–14.3, 18.068–18.168, and 21–21.5 MHz.

_____

_____

15. Turn off the BFO. Tune the 5- to 18-MHz range on a SW radio. Look for foreign broadcast radio stations. The frequencies near 6, 9, and 13 MHz are particularly rich in SW BC stations. Look for as many English-language stations as possible. Record their call letters and locations if possible.

_____

_____

16. Tune for the U.S. time station WWV at 10, 15, or 20 MHz. Listen to this station about 5 min before the hour. Note what happens at the turn of the hour.

_____

_____

Questions

1. What is the main disadvantage of AM over FM for most communications applications? Did you experience it during your monitoring?

2. Name two other AM applications in communications.

3. Explain the need for and operation of the squelch and noise blanker controls.

4. Why is a BFO needed to receive SSB signals?

5. (True or False) CW signals are a form of AM. _____

Chapter

4

Amplitude Modulation and Demodulation Circuits

Project 4-1 Diode Modulator and Mixer

Objective

To demonstrate how a diode can be used to perform amplitude modulation and mixing.

Required

You will use all of the components and test equipment in Project 3-1 plus the following:

1   741 IC op amp
1   1N4149 or 1N4148 silicon diode
1   2-mH inductor
3   0.1-μF capacitors
1   0.01-μF capacitor
4   10-kΩ resistors
1   15-kΩ resistor
1   100-kΩ pot

Introduction

Amplitude modulation and mixing are actually the same process. Two or more signals are combined in a circuit that produces new signals at different frequencies. The amplitude modulator generates sidebands whose frequencies are the sum and difference of the frequencies of the carrier of the modulating signal. A mixer produces similar signals. The output of a mixer consists of the two original input signals plus the sum and difference frequencies, which are, of course, the same as the sidebands.

One of the simplest ways to perform mixing or amplitude modulation is to combine the two signals linearly in a resistive network or op-amp summer and then feed them to a diode rectifier. The output of the rectifier consists of a tuned circuit or filter that selects the desired output frequency. In this experiment, you will demonstrate the use of a diode for amplitude modulation and mixing.

Procedure

Modulator

1. Modify the function generator circuit you used in Project 3-1; the revised circuit is shown in Fig. 4-1. First remove the 10- and 6.8-kΩ resistors connected to pin 1 of the 2206. Also remove the 0.47-μF capacitor and the 1-kΩ pot.

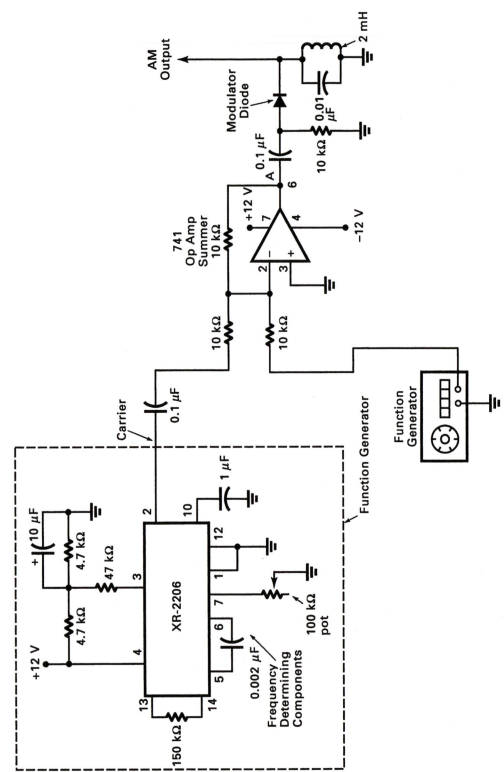

Fig. 4-1 Diode (AM) modulator circuit.

Then connect pin 1 of the 2206 to ground. This modification removes the amplitude modulation capability of the 2206. In this experiment, you will use the function generator simply as a carrier oscillator.

2. Construct the remaining portion of the circuit shown in Fig. 4-1. The circuit consists of an op amp summer used as a linear mixer and a diode modulator mixer circuit. This first circuit will be used for amplitude modulation.

3. Apply power to the circuit. Monitor the output of the 2206 function generator at pin 2. You should observe a sine wave output with an amplitude of approximately 4 V_{p-p}. The input signal from the external function generator should be zero at this time. Next, connect the frequency counter to the function generator output. Then adjust the 100-kΩ potentiometer for an output frequency of 40 kHz. If you do not have a frequency counter, simply use the calibrated horizontal sweep of the oscilloscope to help you in setting the correct frequency.

4. Connect the oscilloscope to pin 6 of the 741 op amp summer. You should observe the 40-kHz signal. Now begin increasing the amplitude of the external signal from the function generator. This is the modulating signal, which can have almost any desired frequency; set it to 1 kHz for this exercise. Adjust the amplitude of the 1-kHz signal for about 5 V_{p-p} at the op amp summer input.

5. Observe the output signal that appears at pin 6 of the 741 op amp summer. Sketch the waveform you obtained. Is this an amplitude-modulated signal? If not, explain what it is.

6. Connect the oscilloscope across the 2-mH inductor. This inductor and the 0.01-μF capacitor in parallel with it form a resonant circuit that will select the desired output signal. In this case, it will pass the carrier and sidebands of an AM signal. Calculate and record the resonant frequency of this tuned circuit.

7. Reduce the amplitude of the modulating signal to zero. With the oscilloscope across the parallel-tuned circuit, adjust the 100-kΩ frequency potentiometer on the 2206 function generator. Tune the frequency for maximum output voltage on the oscilloscope. In doing so, you are adjusting the frequency of the carrier to the resonant frequency of the tuned circuit. Measure and record the output frequency with the frequency counter. How does the measured frequency compare to the frequency you calculated above?

8. While observing the output across the tuned circuit, begin increasing the amplitude of the modulating signal from the external function generator. Set the amplitude of the modulating signal to approximately 1 V_{p-p}. Now observe the signal across the tuned circuit. Is this an AM signal?

9. Continue to increase the amplitude of the modulating signal while observing the effect on the output waveform. Note the effect that it has on the output. Can this circuit achieve 100 percent modulation?

10. You will now modify the circuit and demonstrate its operation as a mixer. First turn off the power. Remove the 2-mH inductor and the associated 0.01-μF capacitor from the output. Replace them with a 15-kΩ resistor in parallel with a 0.1-μF capacitor as shown in Fig. 4-2.

11. Apply power to the circuit. Observe the output of the 2206 function generator with the oscilloscope. Connect the frequency counter to the output of the 2206. Adjust the 100-kΩ pot on the 2206 for a frequency of 20 kHz. This is one input to the mixer.

12. Next, observe the output from the external function generator at the second input to the op amp summer. With the frequency counter, adjust the external function generator output for a frequency of 21 kHz. Set the output amplitude of the external function generator to a value of 1 V_{p-p}. This is the second input to the mixer.

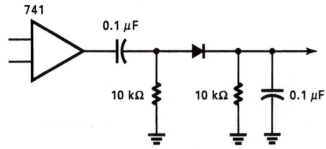

Fig. 4-2 Modifying the diode circuit for mixer operation.

13. From what you have learned about mixers, calculate and record the output frequencies that would occur at the diode mixer output.
14. Considering the components connected to the output of the diode mixer, which of the output signals you predicted in step 13 would appear at the output?
15. Connect the oscilloscope and the frequency counter to the 15-kΩ load. Observe the signal at the output of the mixer. What are its shape and frequency?

Questions

1. What mathematical process is involved in linear mixing?
   a. Addition
   b. Subtraction
   c. Multiplication
   d. Division
2. What mathematical process is involved in amplitude modulation or non-linear mixing?
   a. Addition
   b. Subtraction
   c. Multiplication
   d. Division
3. (True or False) Nonlinear mixing and AM are the same process.
4. What is the function of the 0.1-μF capacitor shown in Fig. 4-2?
   a. Resonant circuit
   b. Noise elimination
   c. High-pass filter
   d. Low-pass filter
5. If the circuit shown in Fig. 4-2 is replaced with the one shown in Fig. 4-3, what output signal will you expect to see?
   a. Difference
   b. Sum
   c. Carrier
   d. Modulating signal

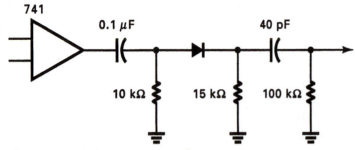

Fig. 4-3 Circuit for question 5.

Project 4-2 Differential Amplifier Modulator

Objective

To demonstrate the use of a differential amplifier as an amplitude modulator.

Required

You will need the 2206 function generator circuit and all of the other components and test equipment listed for Project 3-1. In addition, you will need:

3   NPN transistors (2N4401, 2N4124, 2N3904, MPSA20, etc.)
1   741 op-amp
1   2-mH inductor
3   0.1-μF capacitors
1   0.47-μF capacitor
1   4.7-kΩ resistor
1   6.8-kΩ resistor
2   15-kΩ resistors
1   47-kΩ resistor
1   10-kΩ pot

Introduction

A differential amplifier makes an ideal amplitude modulator for low-level circuits, and a standard circuit of one is shown in Fig. 4-4. The differential transistors Q_1 and Q_2 are connected so their emitters are common. Inputs are applied to the base of one transistor or the bases of both transistors. Two

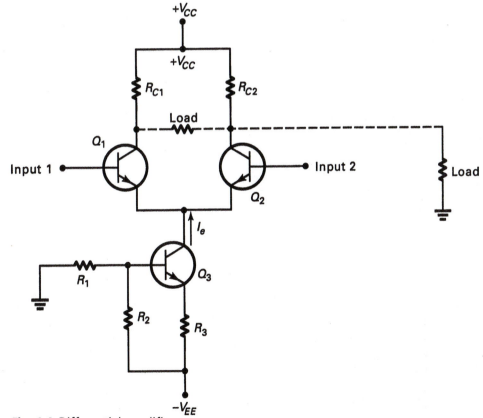

Fig. 4-4 Differential amplifier.

outputs at the transistor collectors are provided. A load can be connected between the two collectors to provide a differential output. Loads can also be connected between the two collectors for balanced operation or between either collector and ground to provide single-ended operation.

The common current I_e to the differential transistors is supplied by a constant-current source, transistor Q_3. Bias resistors R_1 through R_3 and the negative supply voltage V_{EE} set the current value. The emitter current I_e remains constant despite any input circuit variations. The current splits between the two differential transistors. As the base inputs to Q_1 or Q_2 vary, the emitter current divides between the two transistors in proportion to the signal applied. The sum of the currents through Q_1 and Q_2 remains constant and equal to the emitter current.

The approximate single-ended gain A_s of a differential amplifier can be computed with the formula

$$A_s = \frac{R_c}{2.5 \, R_e}$$

This gain is the ratio of the input voltage connected to one of the base inputs and the output voltage taken from either collector to ground. R_c is the value of the collector resistor, and R_e is 25 mV/(I_e/2), where I_e/2 is the current through each differential transistor.

For example, if $R_c = 10 \text{ k}\Omega$ and $I_e/2 = 1.2 \text{ mA}$,

$$R_e = \frac{25 \times 10^{-3}}{1.2 \times 10^{-3}} = 20.83$$

Therefore, the gain is

$$A_s = \frac{R_c}{2.5 \, R_e} = \frac{10,000}{2.5(20.83)}$$

$$= \frac{10,000}{52.1} = 192$$

When the gain formula is rearranged algebraically in terms of I_e, it becomes

$$A_s = \frac{R_c I_e}{125}$$

where I_e is the current from the current source, in milliamperes.

The circuit gain is directly proportional to the emitter current and collector resistance. Under ordinary conditions, both I_e and R_c are fixed values, thereby providing a fixed circuit gain. In amplitude modulation, however, modifications are made to the circuit to cause the emitter current to vary. That is done by applying the modulating or information signal to the constant-current source Q_3, which causes the emitter current to vary linearly and thereby change the emitter current. Changing the emitter current, in turn, changes the circuit gain. By using the differential amplifier to amplify a higher-frequency carrier signal, modulation is achieved. As the modulating signal varies the circuit gain, the output amplitude of the carrier varies proportionally.

Figure 4-5 shows an amplitude modulator made with a differential amplifier. The circuit is connected in a single-ended mode; the carrier is applied to the base of Q_1 through capacitor C_1; and the base of Q_2 is connected directly to ground.

The constant-current source Q_3 is biased into conduction by resistors R_1 through R_3, and the modulating signal is applied to the base of Q_3 through capacitor C_2. The input transistor Q_1 is connected as a follower and has no collector load resistor. Q_2 has a load resistor and serves as the amplifying stage. The output is taken from across the collector load resistor R_5.

As the modulating signal varies, the amplitude of the carrier will vary at the output. The differential amplifier serves as a modulator generating the sidebands and carrier at the output. An appropriate filter or tuned circuit at the output is required to eliminate the low-frequency modulating signal, which

32

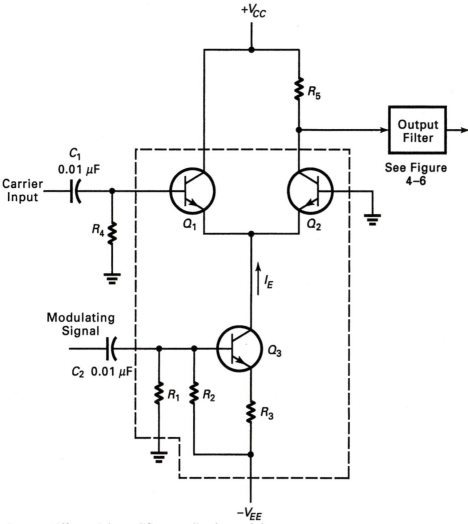

Fig. 4-5 Differential amplifier amplitude modulator.

also appears in the output. The output network can be a simple high-pass filter as shown in Fig. 4-6(*a*) or (*b*) or a tuned circuit as shown in Fig. 4-6(*c*) and (*d*). The output signal consists of the carrier and sidebands.

It is the amplitude of the modulating signal that determines the percent of modulation, and the signal is increased until 100 percent modulation is obtained. The negative-going portion of the modulating signal is capable of reducing the gain of the differential amplifier to zero; the positive-going portions can double the circuit gain. Therefore, the circuit is capable of producing 100 percent modulation. Excessive modulating signal input will cause clipping and distortion as in other amplitude modulators.

In this experiment, you will demonstrate the use of a differential amplifier as an amplitude modulator. Such a circuit is widely used; for example, one version of it produces AM in the 2206 IC function generator you used in the preceding experiment.

Procedure

1. Wire the differential amplifier amplitude modulator circuit shown in Fig. 4-7. Q_1–Q_3 can be any common NPN transistor. The carrier is supplied by the 2206 function generator, and the modulating signal is supplied by an external function generator. Initially, do not connect the 0.1-μF capacitor and the tuned circuit to the collector of Q_3.

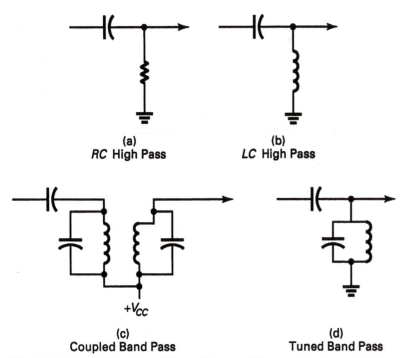

(a)
RC High Pass

(b)
LC High Pass

(c)
Coupled Band Pass

(d)
Tuned Band Pass

Fig. 4-6 Output filters for differential amplifier amplitude modulator.

2. Reduce the output of the external function generator supplying the modulating signal to zero. With an oscilloscope, check at the base of Q_3 to ensure that no ac signal is present.

3. Observe the carrier at the output of the 741 op amp follower. Using the 100-kΩ pot, adjust the frequency of the 2206 to 12 kHz. Apply approximately 100 mV of carrier signal to the modulator at the base of Q_1 by adjusting pot R_4. Observe the differential amplifier output at the collector of Q_2. It should be an undistorted sine wave; if it is not, reduce the input to Q_1 with R_4 until any distortion is eliminated.

4. Connect the 0.1-μF capacitor and tuned circuit to the collector of Q_3 as Fig. 4-7 shows.

5. Now connect the oscilloscope across the tuned output circuit. Vary the 100-kΩ pot on the 2206 function generator to set the carrier frequency to the resonant frequency. Tune for maximum output voltage.

6. Given the values in Fig. 4-7, calculate the resonant frequency f_r. Measure the carrier frequency with a frequency counter.

7. Set the modulating signal frequency to 300 Hz. While observing the carrier output across the tuned circuit, begin increasing the amplitude of the modulating signal from the external function generator. Note the effect on the output signal. You should see a clean amplitude-modulated wave.

8. Adjust the output waveform for 100 percent modulation.

9. Continue to increase the amplitude of the modulating signal to determine if it is possible to overmodulate the differential amplifier and produce distortion and clipping. Is it?

Questions

1. In a differential amplifier, AM is achieved by varying what characteristic of the circuit with the modulating signal?
   a. Bandwidth
   b. Gain
   c. Resonant frequency
   d. Distortion

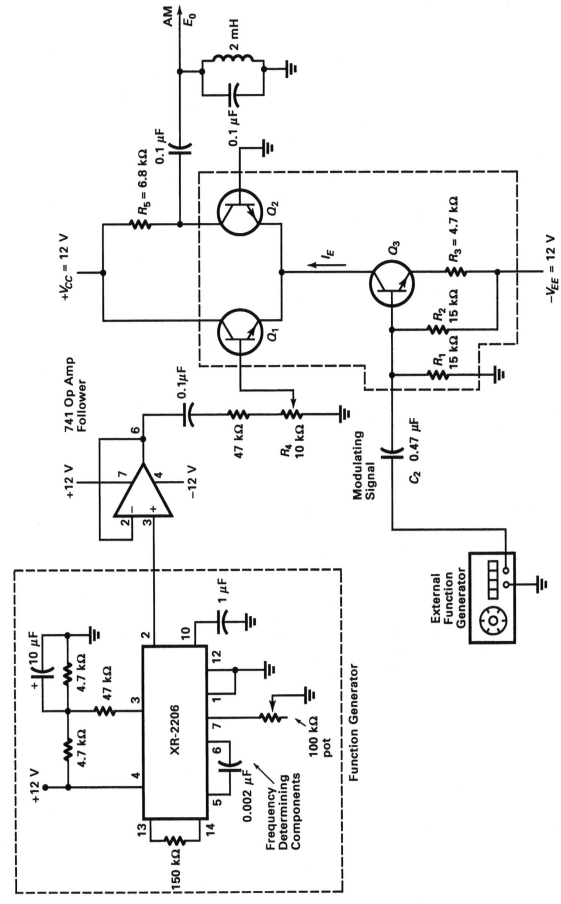

Fig. 4-7 AM with a differential amplifier.

2. The modulating signal causes what parameter in the differential amplifier to vary?
   a. Base voltage of input transistor
   b. Emitter resistance of current source
   c. Collector voltage of Q_1
   d. Emitter current of the differential pair
3. The output tuned circuit is needed to _____.
   a. Increase gain
   b. Suppress the carrier
   c. Eliminate the modulating signal
   d. Eliminate the sidebands
4. (True or False) One-hundred percent modulation can be achieved with the differential amplifier modulator.
5. Assuming I_e is constant, what component in Fig. 4-7 would you change to increase the gain of the circuit?
   a. R_5
   b. R_2
   c. R_4
   d. R_3

Project 4-3 Diode Detector

Objective

To construct a diode detector circuit and demonstrate its use in demodulating an AM signal.

Required

In addition to the equipment and components you used in Project 3-1, you will need the following:

1   741 IC op amp
1   silicon signal diode (1N914, 1N4148, 1N4149, etc.)
1   470-pF capacitor
1   0.005-μF capacitor
2   0.1-μF capacitors
1   1-kΩ resistor
1   10-kΩ resistor
2   100-kΩ resistors
1   10-kΩ pot

Introduction

The process of recovering the original information signal from an AM waveform is known as demodulation. A variety of circuits can be used to demodulate an AM signal, but the one that is simplest and most widely used is the diode detector. A typical circuit is shown in Fig. 4-8. The rectifier diode D_1 removes one-half of the AM signal, and the capacitor C_2 across the load R_2 acts as a low-pass filter to remove the carrier, leaving only the original modulating signal. The waveforms shown in Fig. 4-9 illustrate this process.

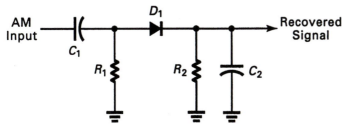

Fig. 4-8 Diode AM detector circuit.

The value of capacitor C_2 is critical in that it determines the degree to which the carrier is filtered out. The capacitor must be large enough to filter the carrier out, but if it is too large, it will cause distortion of the recovered signal. Thus, a compromise value is selected.

In this experiment, you will construct a diode detector circuit, show how it demodulates an AM signal, and demonstrate the effect of different values of filter capacitance on the signal waveshape.

Procedure

1. The modulator circuit you built in Project 3-1 will be used as the AM signal source in this experiment. Rewire the circuit to conform to Fig. 4-10. The modulating signal will come from an external function generator; the 741 op amp amplifies the AM signal to a level suitable for detection; and the 10-kΩ pot is used to adjust the carrier level.

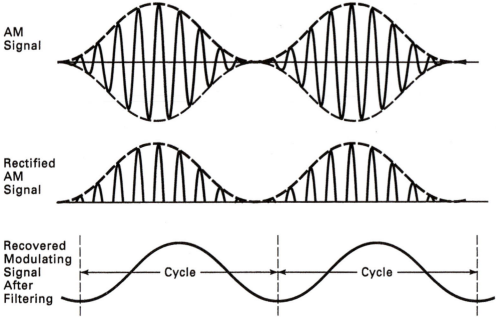

AM
Signal

Rectified
AM
Signal

Recovered
Modulating
Signal
After
Filtering

|← —— Cycle —— →|← —— Cycle —— →|

Fig. 4-9 Demodulation of an AM signal.

2. Next, construct the diode detector circuit shown in Fig. 4-11. Note that a filter capacitor is not connected at this time.

3. Apply power to the circuit. Reduce the modulating signal amplitude to zero. While observing the carrier output of the 741 op amp, adjust the 100-kΩ pot on the 2206 for an output frequency of 30 kHz. Then adjust the 10-kΩ pot for an amplitude of 1.5 V_{p-p}. Set the modulating signal frequency to 200 Hz, and then increase the modulating signal amplitude while observing the op amp output until 100 percent AM is obtained.

4. Observe the diode detector output across the 100-kΩ load resistor. Reverse the diode polarity and note the output waveform. Again reverse the diode connections.

5. Connect a 0.005-μF capacitor across the diode detector load as shown in Fig. 4-12. Observe the diode detector output and sketch the output waveform. If the signal is distorted, reduce the modulating signal amplitude until the distortion is minimized.

6. Connect the 470-pF and the 0.1-μF capacitors, one at a time, in place of the 0.005-μF capacitor as filters in the diode detector circuit. Note the resulting output waveforms.

7. Considering the waveforms you observed in steps 5 and 6, explain how small and large values of capacitance affect the output signal.

8. Reconnect the 0.005-μF capacitor across the 100-kΩ load, and then reverse the connections to the diode. Monitor the output waveform.

9. Compare your results to the result obtained in step 5. Explain the reason for the waveform you observed in step 8.

10. With the oscilloscope, monitor the AM waveform input to the diode detector. Increase the amplitude of the modulating signal to produce overmodulation and clipping.

11. Observe the recovered waveform. Explain the results you obtained.

12. While observing the recovered signal at the output of the diode detector, vary the modulating signal amplitude from zero until distortion occurs. Explain what effect the percent of modulation has on the output signal.

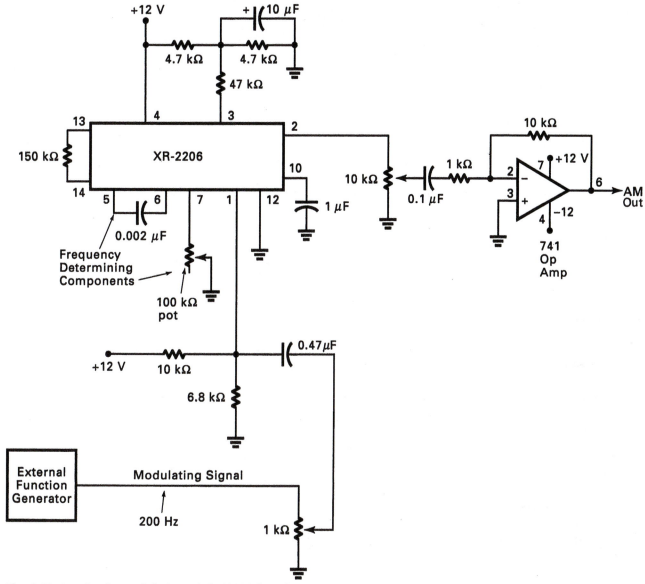

Fig. 4-10 Amplitude modulation of the 2206 function generator IC.

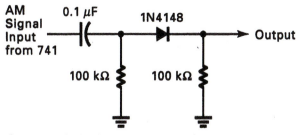

Fig. 4-11 Diode detector without filter capacitor.

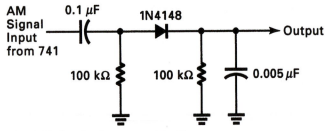

Fig. 4-12 Diode detector with filter.

Questions

1. If the filter capacitor in a diode detector is too small, the _____.
   a. Carrier is not removed
   b. Recovered signal is distorted
   c. Carrier is distorted
   d. Signal amplitude is reduced

2. If the filter capacitor in a diode detector is too large, the _____.
   a. Carrier is not filtered out
   b. Recovered signal is distorted
   c. Carrier is distorted
   d. Signal amplitude is reduced

3. Decreasing the percent of modulation causes the demodulated signal amplitude to _____.
   a. Increase
   b. Decrease
   c. Remain the same
   d. Drop to zero

4. (True or False) Distortion produced by overmodulation is filtered out in the diode detector.

5. Reversing the polarity of the detector diode causes the recovered signal to _____.
   a. Be inverted
   b. Be lowered in amplitude
   c. Be increased in amplitude
   d. Remain the same

40

Project 4-4 Double-Sideband Balanced Modulator

Objective

To demonstrate the operation of a balanced modulator in producing a double-sideband signal.

Required

In addition to the 2206 function generator circuit that you built previously and all of the test instruments used earlier, you will need the following components:

1   741 IC op amp
1   1496 balanced modulator IC
2   100-Ω resistors
1   560-Ω resistor
3   1-kΩ resistors
2   3.3-kΩ resistors
2   4.7-kΩ resistors
1   6.8-kΩ resistor
2   10-kΩ resistors
2   0.1-μF capacitors
1   100-μF electrolytic capacitor
1   10-kΩ pot
1   50-kΩ pot

Introduction

Double-sideband (DSB) modulation refers to amplitude modulation but with the carrier removed. The modulator suppressed the carrier but produces full upper and lower sidebands as may be created by the modulating signal. Single-sideband (SSB) operation refers to amplitude modulation with the carrier suppressed but with one sideband removed. All of the intelligence is carried in either the upper or the lower sideband. SSB operation has the advantage of requiring one-half the spectrum space required by a normal AM or DSB signal.

The key circuit in generating a DSB or SSB signal is the balanced modulator. This circuit produces AM but suppresses the carrier, leaving only the two sidebands in the output. For SSB operation, a highly selective filter is normally used to eliminate one of the sidebands.

A variety of different balanced modulator circuits are used to suppress the carrier. One of the most effective and widely used is the differential amplifier balanced modulator. One such popular integrated circuit is the 1496. It can produce carrier suppression up to 65 dB and has a useful gain up to 100 MHz. It uses specially connected differential amplifiers to cancel the carrier.

In this experiment, you will demonstrate the use of the 1496 to produce a DSB signal. You will also show how this IC can be used to produce conventional amplitude modulation.

Procedure

1. Construct the circuit shown in Fig. 4-13. The 2206 integrated circuit function generator will serve as the carrier source. You will use a 741 op amp inverter to buffer the output signal from the function generator. The output of the 741 op amp becomes the carrier input to the 1496. The modulating signal is supplied by an external function generator. Be particularly careful in constructing this circuit, because there are many components and wiring mistakes are easy to make. Once you have built the circuit, double-check your wiring before applying power.

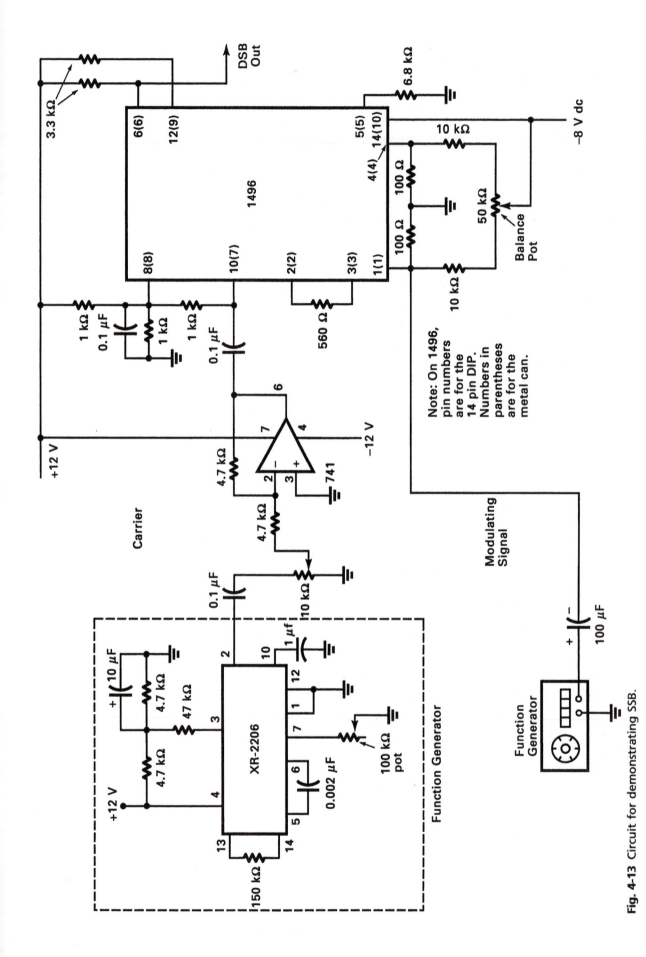

Fig. 4-13 Circuit for demonstrating SSB.

2. While monitoring the carrier signal at the op amp output on pin 6, adjust the 10-kΩ pot for a carrier amplitude of 2 $V_{p\text{-}p}$. By using the 50-kΩ pot on the 2206, set the carrier frequency to 20 kHz, and then adjust the modulating signal input from the external function generator for a frequency of 400 Hz with an amplitude of 0.5 V.

3. Connect the oscilloscope to pin 6 of the 1496 to observe the double-sideband output signal. The classical DSB signal pattern should appear. Adjust the 50-kΩ balance pot on the 1496 so that the arm is approximately in the center of its rotational range. Then fine-tune the pot adjustment until the positive and negative peaks of the DSB waveform are equal.

4. Readjust the 100-kΩ pot on the 2206 function generator to set the carrier frequency to approximately 6 kHz. The adjustment should not change the shape of the DSB output signal, but you should be able to see the carrier cycles more clearly. Next, by using the scope horizontal and vertical gain and triggering controls, expand the DSB signal so you can clearly see the portion of the signal where it goes to zero or reaches a null point between the peaks. You should be able to stabilize the waveform for proper viewing by making slight frequency adjustments with the 50-kΩ carrier-frequency pot. In that way, you will be able to clearly see the phase shift that is characteristic of the DSB signal. (Refer to Figs. 3-15 and 4-27 in the text.) This is one way to clearly identify a DSB signal from a conventional AM signal.

5. Readjust the carrier frequency to approximately 20 kHz and again display the DSB signal, and then rotate the balance pot in one direction or the other until a conventional AM signal is obtained. If the AM signal does not appear when you are adjusting the pot in one direction, adjust the pot in the opposite direction until it does. You may have to reduce the amplitude of the modulating signal from the external function generator as well to make the AM signal appear. Then, by adjusting the modulating signal input level, you should be able to obtain any percent of modulating desired. By varying the 100-kΩ pot, you have unbalanced the circuit and thereby allowed the carrier to be inserted. With the pot in the center of its rotation, the internal IC circuitry is carefully balanced so the carrier is suppressed. In the unbalanced condition, the carrier passes through, causing traditional amplitude modulation to occur.

Questions

1. The circuit that produces amplitude modulation but suppresses the carrier is known as a(n) _____.
   a. Op amp
   b. Function generator
   c. Balanced modulator
   d. Differential amplifier
2. A DSB signal is made up of the sum of the _____.
   a. Upper and lower sidebands
   b. Upper and lower sidebands and the carrier
   c. Carrier and the upper sideband
   d. Carrier and the lower sideband
3. (True or False) If properly adjusted, an IC balanced modulator can also be used to produce AM.

Chapter

5

Fundamentals of Frequency Modulation

Project 5-1 Frequency Modulation

Objective

To demonstrate frequency modulation with a voltage-controlled oscillator.

Required

You will need the 2206 function generator circuit you constructed previously plus an oscilloscope, a frequency counter, and an external function generator. You will also need a 10-μF capacitor and a 100-kΩ resistor.

Introduction

In frequency modulation, the modulating signal varies the frequency of the carrier. The carrier amplitude remains constant, but the increasing and decreasing amplitude of the modulating signal causes the carrier to deviate from its center frequency. The amount of deviation is a function of the amplitude of the modulating signal. The rate of deviation is proportional to the frequency of the modulating signal.

There are many different ways to produce frequency modulation. Today, one of the easiest ways is simply to apply a modulating signal to a voltage-controlled oscillator (VCO), and numerous IC VCOs are available. The 2206 function generator IC you have used in past experiments uses a VCO as the primary signal-generating circuit. This oscillator can easily be frequency-modulated simply by applying the appropriate modulating signal. In this experiment, you will demonstrate frequency modulation with the 2206.

Procedure

1. The circuit for this experiment is shown in Fig. 5-1. This is the same 2206 function generator you have used previously, but note that you will add a 10-μF capacitor and a 100-kΩ resistor to pin 7. To the capacitor you will connect an external function generator to serve as the modulating signal.
2. Apply power to the circuit. Reduce the modulating signal amplitude from the external function generator to zero, and then observe the carrier output at pin 2 of the 2206. Adjust the 100-kΩ pot on the 2206 for a frequency of approximately 30 kHz. Set the horizontal sweep controls on the oscilloscope to display approximately three cycles of the carrier wave.
3. Slowly increase the amplitude of the modulating signal from the external function generator and set the frequency of the signal to approximately 200 Hz. As you increase the modulating signal amplitude, note the display on the oscilloscope. You should see the carrier wave begin to "vibrate," which indicates a change in frequency with the modulating signal.

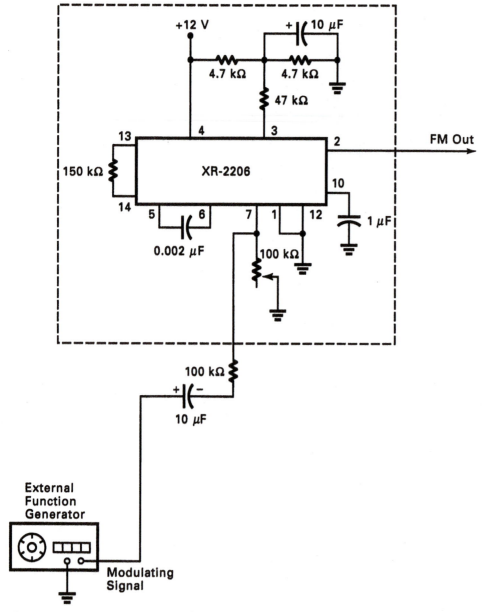

Fig. 5-1 Frequency modulation of the 2206.

You should observe a waveform that looks approximately like the one shown in Fig. 5-2. Continue slowly increasing the modulating signal amplitude and note the effect on the carrier wave. If you have difficulty obtaining a stable display, use the triggered sweep function on the oscilloscope. Manipulate the trigger and horizontal frequency controls until a stable waveform is obtained. What you are seeing is the carrier frequency being instantaneously changed by the modulating signal. While varying the amplitude of the modulating signal and observing the resulting FM output, determine the relationship between the frequency deviation and the modulating signal amplitude.

4. Set the amplitude of the modulating signal at the 10-μF capacitor to 2 $V_{p\text{-}p}$, and then observe the carrier output on the oscilloscope. You will see the waveform shown in Fig. 5-2, indicating frequency deviation. Now, while observing the waveform, vary the frequency of the modulating signal. Note the effect on the waveform. How does the deviation change with respect to the modulating signal frequency?

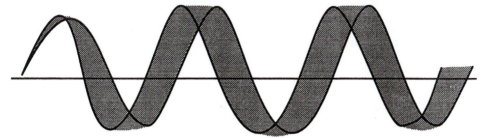

Fig. 5-2 How FM looks on an oscilloscope.

Questions

1. Which of the following is the most correct statement with regard to the relationship between the carrier frequency deviation and the amplitude of the modulating signal?
   a. Decreasing the modulating signal amplitude increases the carrier deviation.
   b. Decreasing the modulating signal amplitude decreases carrier deviation.
   c. Carrier deviation is not affected by the amplitude of the modulating signal.
   d. The frequency deviation varies in direct proportion to the modulating signal frequency.

2. Varying the frequency of the modulating signal causes the frequency deviation of the carrier to _____ .
   a. Increase
   b. Decrease
   c. Remain the same
   d. Drop to zero

3. The type of modulation produced by the VCO in the 2206 IC is _____ .
   a. Frequency modulation
   b. Phase modulation
   c. Indirect FM
   d. FSK

4. What is the deviation ratio of an FM system in which the maximum permitted frequency deviation is 10 kHz and the maximum modulating frequency is 3 kHz?
   a. 0.3
   b. 1
   c. 3
   d. 3.33

5. The number of sidebands produced by a sine-wave carrier being frequency-modulated by a single-frequency sine-wave tone is _____ .
   a. 1
   b. 2
   c. 4
   d. Infinite

Project 5-2 FM Radio Experience

Objective

To tune and listen to radio stations using frequency modulation.

Required

1   FM broadcast radio
1   receiver capable of receiving two-way FM communications in the 140- to 170-MHz and 420- to 470-MHz ranges. (A common scanner radio with a frequency range of 50 to 1300 MHz is the best way to listen to all of the signals in this range.)
2   family radio units
•   Antennas as required

Introduction

Most two-way radio communications use FM. These include police, fire, taxi, public service, marine, and many others. Most of this communication takes place in the VHF range from 150 to 170 MHz. The 420- to 470-MHz frequency range is also widely used. You can hear most of the conversations in these ranges using a scanner receiver. In this experiment, you will experiment with a scanner receiver and use it to listen to as many different types of FM communications as possible.

Procedure

1. Read the instruction manual for the scanner receiver.
2. State the frequency range of the FM broadcast band and the frequency spacing of stations.
3. Use an FM radio to listen to the broadcast band. Tune the entire extent of the band and count the number of stations you hear and record their frequency. Describe what you hear.
4. Set the scanner receiver to receive stations in the 152- to 155-MHz range. You should hear a variety of signals. Describe what you hear, including the source of the transmissions.
5. Try out the squelch control on the receiver.
6. Monitor the channels in the 156- to 158-MHz range. Describe the type of stations you hear, if any.
7. Listen to stations in 162- to 163-MHz range. What do you hear?
8. Set the scanner to listen to stations in the 146- to 148-MHz range. Again, describe what you hear.
9. Set the scanner to scan the 450- to 470-MHz range. Describe some of the conversations you hear. What are the sources?
10. Scan the 46- to 47-MHz and 49- to 50-MHz ranges. What do you hear, if anything? What should you hear?
11. Try out the family radio units. Can you change frequencies? Explain. List all controls on the unit.
12. Experiment with the maximum communications range, both indoors and outdoors. Record your results and explain.

Question

1. What is the main advantage of FM over AM for most communications applications?

_____

_____

Chapter 6

FM Circuits

Project 6-1 Phase-Locked Loops

Objective

Review the concepts and operation of a phase-locked loop.

Introduction

The phase-locked loop (PLL) has become one of the most widely used circuits in communications. It can operate as a filter, FM demodulator, frequency synthesizer, clock frequency multiplier, and clock recovery circuit. This electronic feedback control circuit is one of the more fascinating and useful of all communications circuits. In this project, you will review this important circuit.

Required

PC with Internet access.

Procedure

1. Go to the website www.work-readyelectronics.org.
2. At the top of the page, click on the Modules box, which will take you to a second page listing all the various modules you can access for free.
3. Answer the three questions given, then select the Phase-Locked Loop module.
4. When you have accessed the module, follow the instructions given and complete the module.
5. Be sure to take advantage of the Knowledge Probes that help you to review the material.
6. Take the formal Assessment at the end of the module.
7. While accessing the Phase-Locked Loop module, go to the Learning Resources section and select the PLL Circuit Lab experiment. This hardware experiment can be run next, or proceed to Project 6-2 in the manual as a similar alternative.

In this experiment, you will demonstrate PLL operation and measure the phase shift between the input signal and the VCO during the locked state. You will also demonstrate how the VCO output tracks a variable-frequency input. Finally, you will calculate and measure the free-running frequency and the lock range of the 565 PLL.

Procedure

1. Connect the circuit shown in Fig. 6-2. An external function generator with variable-frequency capability is connected as the PLL input. Reduce the amplitude to zero for now.
2. Use the formula for the free-running frequency of the VCO, given earlier, to calculate that frequency by using the values of R_1 and C_1 in Fig. 6-2.
3. Apply power to the circuit shown in Fig. 6-2. By using a frequency counter, measure the output frequency of the VCO at pin 5 on the 565 IC. Observe the output signal with an oscilloscope.
4. Compare your computed and measured values of free-running frequency and explain any difference.
5. Adjust the function generator input to the PLL input for an amplitude of 1 V each peak to peak of sine wave. Set the external function generator for a frequency approximately equal to the free-running frequency of the VCO.
6. Use an oscilloscope and/or frequency counter to measure the function generator output frequency and the VCO output frequency at pin 5. Are the two frequencies the same? Why or why not?
7. Use a dual-trace oscilloscope to display the function generator and VCO output signals simultaneously. Measure the amount of phase shift between the two. You can do that by measuring the time shift t between corresponding parts of the two waveforms and then using the formula

$$\text{Phase shift, in degrees} = \frac{360t}{T}$$

where T is the period of the signals.

8. While continuing to observe the function generator and VCO output signals on the dual-trace oscilloscope, vary the function generator frequency above and below the free-running value. Note the effect on the VCO output. Describe the relationship between the function generator input and the VCO output signal frequencies.
9. Use the formula given earlier to compute the lock range for the PLL. Record your value.
10. Assuming that the lock range is centered on the free-running frequency of the VCO, calculate the upper and lower lock range limits. Is the difference between the upper and lower lock values equal to the lock range?
11. By using the circuit shown in Fig. 6-2, you will now measure the lock range. First, however, set the function generator input frequency to the PLL free-running frequency to ensure initial lock.
12. Begin decreasing the function generator input frequency while observing the VCO output. At some point, you will notice a frequency variation or jitter. Stop decreasing the frequency exactly at that point and measure the function generator frequency. It is the lower lock limit of the PLL. Record it.
13. Increase the input frequency while observing the VCO output. Lock will occur. Keep increasing the frequency. Again you will reach a point where the VCO output begins to jitter. Reduce the input frequency to just below the point at which the jitter ceases. Then measure the function generator output frequency. This is the upper lock limit of the PLL. Record it.
14. By using the experimental data you collected in steps 11 through 13, calculate and record the measured lock range. How does it compare to your calculated value?

52

Questions

1. Varying the PLL input frequency causes the PLL output to _____.
   a. Track the input
   b. Remain constant at the free-running frequency
   c. Vary inversely
   d. Drop to zero

2. The error signal is generated by what characteristic of the input and VCO signals?
   a. Frequency
   b. Phase shift
   c. Amplitude difference
   d. Rise time

3. What is the relationship between the capture f_c and lock f_L ranges of the PLL?
   a. $f_c = f_L$
   b. $f_c > f_L$
   c. $f_c < f_L$
   d. $f_c = 2f_L$

4. To increase the free-running or center frequency of the PLL, what changes should be made in R_1 and/or C_1?
   a. Increase R_1
   b. Increase C_1
   c. Decrease R_1
   d. Decrease C_1
   e. Both a and b
   f. Both c and d

5. If the input frequency to the PLL is outside the capture and lock ranges, the VCO output is the _____.
   a. Upper lock frequency
   b. Free-running frequency
   c. Lower lock frequency
   d. Input frequency

Project 6-3 Frequency Demodulation with a Phase-Locked Loop

Objective

To demonstrate the operation of a phase-locked loop as a demodulator for FM signals.

Required

In addition to an oscilloscope, frequency counter, function generator, and power supply, you will need the 2206 FM modulator you built in Project 5-1 and the 565 PLL circuit you used in Project 6-1. You will also need a 0.001-μF capacitor and a 10-kΩ pot.

Introduction

Perhaps the best demodulator circuit for FM is the phase-locked loop. Because of its excellent fidelity, wide range response, and noise suppression, the PLL produces a clean and faithful reproduction of the original modulating signal. Because PLLs are available in low-cost IC form, they are widely used for this purpose, particularly in critical applications.

To use the PLL as a frequency demodulator, you apply the FM signal to the phase detector input. A recovered signal is taken from the output of the loop low-pass filter (LPF). See Fig. 6-3.

As the carrier varies in accordance with the modulating signal, it produces a varying phase when compared to the VCO. The error signal developed then forces the VCO output to follow the FM input. Remember that, in order for the PLL to remain locked, the two inputs to the phase detector must be the same. The VCO output will track the FM input.

In order for the VCO to duplicate the input FM signal exactly, the signal applied to the VCO control input from the loop filter must be the same as the modulating signal. The output of the loop filter is therefore the recovered intelligence signal.

In this experiment, you will use the 565 IC PLL to demonstrate FM demodulation.

Procedure

1. The complete circuit for this experiment is shown in Fig. 6-4. It is made up of the 2206 IC frequency modulator you used in the preceding experiment, and the demodulator circuit itself is the 565 PLL you used in that experiment. Connect the output of the frequency modulator to the 565. In the PLL circuit, replace the 0.01-μF capacitor with a 0.001-μF capacitor. Replace the 2.7-kΩ resistor R_1 with a 10-kΩ pot.

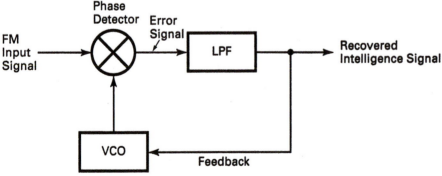

Fig. 6-3 Phase-locked loop used as an FM demodulator.

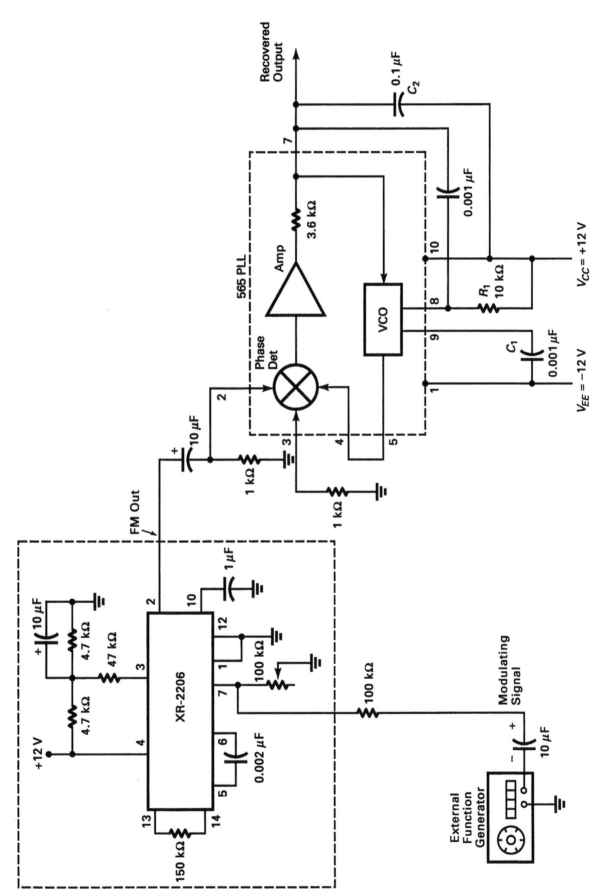

Fig. 6-4 Using a PLL for FM demodulation.

2. Connect a frequency counter and oscilloscope to the 2206 output at pin 2. Reduce the modulating input signal to zero. Adjust the 100-kΩ pot for a carrier frequency of 30 kHz.

3. Connect the frequency counter and oscilloscope to the PLL output at pin 5 on the 565. Adjust the 10-kΩ pot for a frequency of 30 kHz. At this time, the 565 PLL should be locked to the 2206 output; check that by observing the 565 output at pin 5 while varying the 100-kΩ pot on the 2206. The output should track. Readjust the 100-kΩ pot for an output of 30 kHz.

4. Apply a modulating signal from the external function generator to the 2206 FM generator. Use a frequency of 200 to 400 Hz. While observing the output of the 2206 at pin 2, increase the amplitude of the modulating signal until you obtain an FM signal.

5. Observe the PLL loop filter output at pin 7 to see the recovered signal. Compare that signal to the modulating signal applied to the 2206.

6. While observing the recovered output at pin 7 on the 565 PLL, vary both the frequency and amplitude of the modulating signal. Note the effect on the recovered output.

Questions

1. The recovered output from a PLL demodulator appears at the output of the _____.
   a. Phase detector
   b. Loop filter
   c. VCO
   d. *RC* timing network
2. (True or False) The PLL VCO output signal tracks the FM input.
3. The signal controlling the VCO in the PLL is the _____.
   a. FM signal
   b. Carrier
   c. Phase detector output
   d. Original modulating signal
4. If the modulating signal is removed from the modulator circuit, the loop filter output of the PLL will be _____.
   a. The carrier
   b. The modulating signal
   c. The free-running VCO frequency
   d. Zero
5. To what frequency should the PLL VCO free-running be set when the demodulator is used in a superheterodyne receiver?
   a. The input carrier frequency
   b. The manufacturer's recommended frequency
   c. The receiver intermediate frequency
   d. Any frequency higher than the highest-frequency modulating signal

Project 6-4 Pulse-Averaging Discriminator

Objective

To demonstrate the operation of a pulse-averaging discriminator in demodulating FM signals.

Required

You will need the 2206 FM generator you used in Project 5-1 plus the oscilloscope and frequency counter. In addition, you will need the following parts:

1 74123 or 74LS123 dual TTL one-shot (The 74121 and 74122 TTL ICs also will work in this application)
1 0.01-μF capacitor
2 0.02-μF capacitors
1 1-kΩ resistor
1 4.7-kΩ resistor
2 15-kΩ resistors
• Power supplies of +12 V and +5 V

Introduction

There are numerous types of FM demodulator circuits that use the principle of averaging pulses in a low-pass filter to recover the intelligence signal. For example, the quadrature FM detector produces output pulses of varying widths that, when averaged in a low-pass filter, produce the original modulating signal. The pulse-averaging discriminator generates fixed-amplitude, fixed-width pulses, one per cycle of the FM signal. These pulses, when averaged in a low-pass filter, reproduce the original intelligence.

In this experiment, you will demonstrate that concept by constructing a pulse-averaging discriminator made up of a one-shot multivibrator. When triggered by the FM signal, the one-shot, or monostable, produces one fixed-width pulse per cycle. The pulses are fed to a low-pass filter to reproduce the original signal.

Procedure

1. Construct the experimental circuit shown in Fig. 6-5. The output from the 2206 function generator is taken from pin 11; it is a 5-V square wave compatible with the TTL logic circuitry in the 74123 one-shot. The signal is also frequency-modulated and follows exactly the sine-wave output that appears at pin 2 on the 2206.
2. Connect the frequency counter to the 2206 output at pin 11, and then adjust the 100-kΩ pot for a center frequency of 20 kHz. While you do that, be sure to reduce the modulating signal input from the external function generator to zero.
3. Connect the oscilloscope output to pin 13 of the 74123 one-shot. You should observe a chain of pulses, and the pulse width should be approximately 15 μs. By using the 100-kΩ pot on the 2206, vary the frequency over a narrow range above and below the 20-kHz point. Note that the frequency of the pulses changes but the pulse width remains constant.
4. Again set the 2206 function generator for a center frequency of 20 kHz with the 100-kΩ pot, and then apply the modulating signal to the 2206 from the external function generator. Set the modulating signal frequency to 200 Hz and monitor the output of the 2206 at pin 11. You should see a frequency-modulated square wave. This is the signal applied to the one-shot.

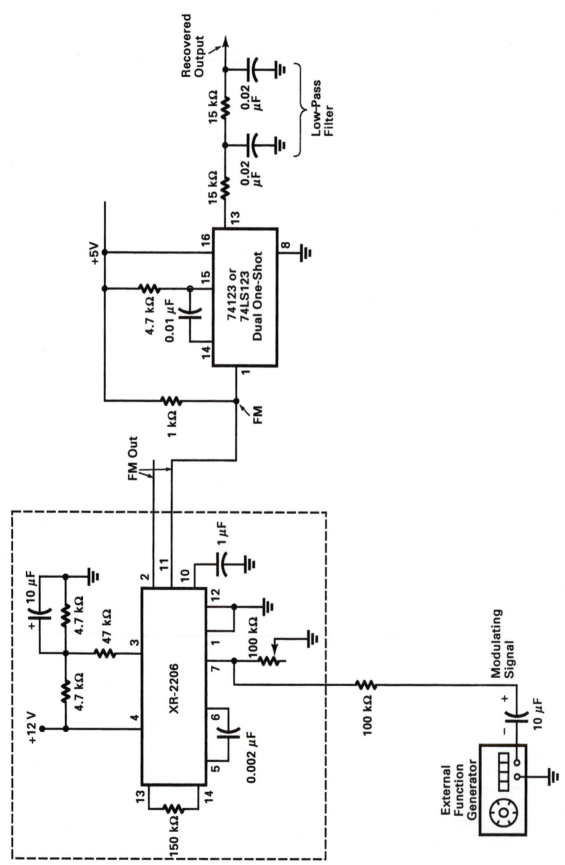

Fig. 6-5 Pulse-averaging discriminator or demodulator.

5. Connect the oscilloscope to the output of the low-pass filter. You should observe the recovered 200-Hz sine wave. Vary the amplitude of the modulating signal and note the effect on the output.

6. While observing the demodulating output, increase the modulating signal frequency from 200 Hz slowly up to approximately 2 kHz. Note the effect on the output. How does the demodulated signal vary with frequency? Explain the reason for the action you observe.

Questions

1. The circuit that generates fixed-width fixed-amplitude pulses is known as a(n) _____.
   a. Astable multivibrator
   b. Pulse generator
   c. Oscillator
   d. One-shot multivibrator

2. The circuit that smooths the pulses into the original intelligence signal is the _____.
   a. Low-pass filter
   b. High-pass filter
   c. One-shot multivibrator
   d. Quadrature detector

3. Another FM demodulator that uses the principle of averaging pulses in a filter is the _____.
   a. Phase-locked loop
   b. Foster-Seeley discriminator
   c. Quadrature detector
   d. Ratio detector

4. The amplitude of the recovered intelligence signal is directly proportional to _____.
   a. Carrier frequency
   b. Carrier deviation
   c. Pulse amplitude
   d. Pulse width

5. As the frequency of the modulating signal increases, the recovered signal output amplitude _____.
   a. Decreases
   b. Increases
   c. Remains the same
   d. Drops to zero

Chapter 7

Digital Communication Techniques

Project 7-1 Digital-to-Analog Conversion

Objective

Review DAC operation and specifications.

Introduction

With most communications techniques being digital today, data conversion techniques are a key part of most wireless systems. There are two data conversion processes that you should understand: analog-to-digital conversion (ADC) and digital-to-analog conversion (DAC). In this project, you will begin a review of these techniques beginning with DAC in the project.

Required

PC with Internet access.

Procedure

1. Go to the website www.work-readyelectronics.org.
2. At the top of the page, click on the Modules box, which will take you to a second page listing all the various modules you can access for free.
3. Answer the three questions given, then select the Data Conversion I module.
4. When you have accessed the module, follow the instructions given and complete the module.
5. Be sure to take advantage of the Knowledge Probes that help you to review the material.
6. Take the formal Assessment at the end of the module.
7. While still accessing the module, go to the Learning Resources section and select lab experiment Digital-to-Analog Converters. Run that experiment as directed.

Project 7-2 Analog-to-Digital Conversion

Objective

Review ADC operation and specifications.

Introduction

This project continues with your review of data conversion techniques that are a key part of most wireless systems today. This project introduces you to the concept of analog-to-digital conversion (ADC) and the most popular types of ADC circuits. Be sure to complete Project 7-1 before working on this project; it is a prerequisite.

Required

PC with Internet access.

Procedure

1. Go to the website www.work-readyelectronics.org.
2. At the top of the page, click on the Modules box, which will take you to a second page listing all the various modules you can access for free.
3. Answer the three questions given, then select the Data Conversion II module.
4. When you have accessed the module, follow the instructions given and complete the module.
5. Be sure to take advantage of the Knowledge Probes that help you to review the material.
6. Take the formal Assessment at the end of the module.
7. While still accessing the module, go to the Learning Resources section and select Lab 1 experiment, Analog-to-Digital Converters. Run that experiment as directed.
8. While still accessing the module, go to the Learning Resources section and select Lab 2 experiment, Analog-to-Digital Converters. Run that experiment as directed.

Project 7-3 Digital Signal Processing

Objective

Review DSP concepts and processes.

Introduction

The heart of most digital communications techniques is digital signal processing (DSP). DSP is the technique of performing signal processing on analog signals by digitizing the signals, then using a variety of mathematical algorithms to perform filtering, compression, equalization, modulation, demodulation, mixing, and other operations. Most of these operations are carried out by a special architecture microprocessor designed for DSP. This module reviews the DSP concepts you learned in the text.

Required

PC with Internet access.

Procedure

1. Go to the website www.work-readyelectronics.org.
2. At the top of the page, click on the Modules box, which will take you to a second page listing all the various modules you can access for free.
3. Answer the three questions given, then select the Digital Signal Processing module.
4. When you have accessed the module, follow the instructions given and complete the module.
5. Be sure to take advantage of the Knowledge Probes that help you to review the material.
6. Take the formal Assessment at the end of the module.

Project 7-4 DSP Tutorial

Objective

Expand your knowledge of DSP principles and techniques.

Introduction

Digital signal processing is now the dominant method of signal processing in communications systems. It is easy to implement in software with a special programmable digital signal processor or, in some cases, a field-programmable logic array (FPGA). The textbook provides only the basics of DSP with an introduction to the major applications such as filtering and spectrum analysis. This activity provides a way to explore DSP in greater depth.

Required

PC with Internet access.

Procedure

1. For a more in-depth look at DSP with minimal math, acquire the book *Digital Signal Processing, A Hands-On Approach* by Charles Schuler and Mahesh Chugani, McGraw Hill, 2005. Be sure to explore the activities on the accompanying CD.
2. Seek out and take any available online tutorials on DSP. Some suggestions:
   a. www.dsptutor.freeuk.com. Take their tutorials on digital filters and look at the Java applets on the fast Fourier transform and aliasing.
   b. www.dspgure.com. Explore the available tutorials and review them.
   c. www.bores.com/courses/intro. Take the Introduction to DSP course.
   d. Search for other DSP tutorials via Google or Yahoo.
3. Search for DSP fundamentals information on the websites of the major DSP chip providers:
   a. Texas Instruments (www.ti.com)
   b. Analog Devices (www.analog.com)
   c. Freescale Semiconductor (www.freescale.com)
   d. Xilinx (www.xilinx.com)
   e. Altera (www.altera.com)

Questions

1. What is the most common solution for the aliasing problem?
2. Name the two most common DSP filters and explain the pros/cons and applications of each.
3. Name two common applications of FFT.
4. What is the fastest DSP chip available (clock speed)?
5. What is the main product of Altera and Xilinx?

Chapter

8

Radio Transmitters

Project 8-1 Class C Amplifiers and Frequency Multipliers

Objective

To demonstrate the operation and use as a frequency multiplier of a class C amplifier.

Required

1 function generator
1 oscilloscope
1 frequency counter
1 digital multimeter
1 NPN transistor (2N3904, 2N4124, 2N4401, 2N2222, etc.)
1 2-mH inductor
1 0.02-μF capacitor
1 0.1-μF capacitor
1 10-kΩ resistor
1 power supply for +5 V

Introduction

Some radio transmitters use class C amplifiers for power amplification, and that is particularly true of FM transmitters. Most AM transmitters or those transmitting SSB use linear amplifiers. Class C amplifiers are usually preferred because they are far more efficient than linears, but they distort the AM/SSB signals. They can be used in FM transmitters, however, which means that they generate more output power vs. input power than any other kind of amplifier.

A transistor class C amplifier is one that conducts for less than one half-cycle (180°) of the input sine wave signal. Class C amplifiers conduct only during the half-cycle when the emitter-base junction in the transistor is forward-biased. Most class C amplifiers use some form of base bias that keeps the transistor cut off until the input signal polarity and amplitude are high enough to cause it to conduct. Typically, the bias is set up so that conduction occurs during the 90° to 150° range. Since conduction is less than one half-cycle, the collector current is, of course, highly distorted. However, since a tuned circuit is connected in series with the collector, the output signal is a complete sine wave. The flywheel effect of the tuned circuit supplies the missing half-cycle.

Class C amplifiers are also used as frequency multipliers. By connecting a tuned circuit in the collector whose resonant frequency is some integral value of the input frequency, frequency multiplication occurs. That provides a simple and low-cost way to increase the signal frequency. In this way, stable low-frequency circuits can be used to generate the original signal, and multipliers are used to increase the signal to its final, higher RF value.

The class C amplifier is a good frequency multiplier because of the distortion it generates. The distortion, of course, means that the output signal contains many harmonics. The tuned circuit in the collector is used to select the desired harmonic.

In this experiment, you will demonstrate a simple transistor class C amplifier and show its use as a multiplier.

Procedure

1. Construct the class C amplifier circuit shown in Fig. 8-1. Initially connect the function generator but be sure that the input amplitude is reduced to zero initially.
2. Observe the tuned circuit in Fig. 8-1 and calculate the resonant frequency by using the values given. Record the frequency.
3. Connect an oscilloscope to the class C amplifier collector output. While observing the signal there, begin increasing the function generator input signal. Apply approximately 1.5 V_{p-p} to the input capacitor. The function generator frequency should be set to the resonant frequency value you computed in step 2. While observing the output, carefully tune the function generator frequency for maximum output voltage at the collector.
4. Measure and record the output frequency with the frequency counter. Your measured value should be close to the computed resonant frequency.
5. While observing the collector output on the oscilloscope, vary the function generator input voltage from zero until the output signal increases. At just the point where the output signal amplitude reaches its maximum output, stop increasing the input voltage.
6. By using a digital multimeter, measure the voltage across the 10-kΩ base resistor. Record the polarity and voltage.
7. While monitoring the dc voltage across the 10-kΩ resistor, continue increasing the input voltage from the function generator. Increase it to a value of about 4 V_{p-p} and note how the bias voltage changes. Does it increase, decrease, remain the same, or otherwise change?

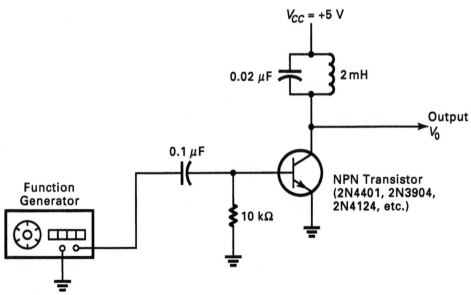

Fig. 8-1 Class C amplifier.

8. Again reduce the function generator input voltage to zero and monitor the collector output. Begin increasing the input voltage slowly and note the point where the output voltage rises to its maximum point. At that time, measure the input voltage. Explain why it is necessary for the input voltage to reach that level before the output signal appears.

9. While observing the output, continue increasing the input and note any change in the output. Measure and record the output voltage. Compare the output voltage you measured to the supply voltage V_{cc} and explain how the output value is obtained.

10. Reduce the input from the function generator to the point where the output signal at the collector just begins to drop off its maximum value. Then insert an ammeter in series with the collector supply. Measure and record the collector current.

11. Calculate and record the input power of this class C amplifier.

12. Assume now that the class C amplifier will be a frequency multiplier. The output will be the resonant frequency of the tuned circuit. Calculate all of the input frequencies of which the resonant frequency will be a harmonic. Calculate and list all the frequencies from the second through the tenth harmonics.

13. While observing the collector output, adjust the function generator input level to just where the output voltage reaches its maximum amplitude. The output level should be approximately 10 V_{p-p}.

14. Now adjust the function generator input frequency to each of the input frequencies you calculated above. Tune the generator carefully until the output signal peaks, and then measure the input frequency with the frequency counter. Repeat that process for all of the harmonic values you calculated in step 12.

15. As you continue to lower the input frequency and produce higher and higher harmonics, note the effect on the output voltage. Does the output increase or decrease as the harmonic number increases?

Questions

1. The type of bias used on the class C amplifier shown in Fig. 8-1 is known as _____ .
   a. Self-bias
   b. External bias
   c. Signal bias
   d. Variable bias

2. If the transistor in the class C amplifier were a PNP, the base bias with respect to ground would be _____ .
   a. Positive
   b. Negative
   c. Zero
   d. Insufficient information is given to answer this question

3. In a properly adjusted class C amplifier, the collector output voltage is _____ .
   a. V_{cc}
   b. $2V_{cc}$
   c. $V_{cc}/2$
   d. A value that depends on the input signal amplitude

4. As the frequency multiplication factor increases, the output signal voltage and power _____ .
   a. Decrease
   b. Increase

R1 biases the transistor into the linear region for maximum gain and least distortion. The frequency of the circuit is set by the crystal that uses its parallel resonant mode. The output, taken from the emitter, is a low-power sine wave.

While the frequency of the output is set by the crystal, any capacitance appearing in series or parallel with the crystal will also affect the frequency. However, a variable capacitor like that shown in Fig. 8-2 is often connected in parallel with the crystal to permit some variation in the output frequency.

Procedure

1. Construct the circuit shown in Fig. 8-2. If you are using a breadboarding socket, be sure to keep the component leads as short as possible to ensure best performance. Initially, *do not* connect the 47-pF capacitor C3 or the 1-kΩ pot. Apply +10 V dc to the circuit. Connect an oscilloscope to the output and observe the signal.
    a. What is the shape of the signal? Is it what you could call distorted?
    b. What is the peak-to-peak output voltage? The rms output voltage?
2. Connect a frequency counter to the output. Set the counter for the best resolution of measurement. Measure the output frequency. How does it compare to the frequency marked on the case of the crystal?
3. Compute difference between the marked and measured frequencies. Express this as a percentage of the marked frequency.
4. Connect the 47-pF capacitor across the crystal as shown in Fig. 8-2. How did the frequency change? Explain.
5. Disconnect the 47-pF capacitor.
6. Again measure the no-load output voltage.
7. Connect the 1-kΩ pot load across the output as shown in Fig. 8-2. Set the pot for maximum resistance of 1 kΩ. Measure the output voltage. How did the voltage change? Explain.
8. Adjust the pot until the output voltage drops to a value exactly one-half of the output voltage with no load. Record that value.
9. Turn off the power. Disconnect the pot from the circuit. Measure the resistance of the pot with an ohmmeter. What is the significance of this value?
10. Using the values you determined in steps 8 and 9, answer these questions:
    a. What is the output power across the load with the pot set as in step 8?
    b. What is the output impedance of this circuit?

Questions

1. (True or False) The frequency of a crystal oscillator can be changed with an external capacitor.
2. The crystal oscillator in this experiment operates in which class?
    a. A
    b. AB
    c. B
    d. C
    e. E
3. A distorted output means that:
    a. The circuit is not operating correctly.
    b. The circuit is operating off frequency.
    c. Harmonics are contained in the output.
    d. Output power will be degraded.
4. The circuit in Fig. 8-2 operates with a
    a. Common emitter
    b. Common collector
    c. Common base
5. The circuit in this experiment uses which mode of crystal operation?
    a. Parallel
    b. Series

Project 8-3 Building and Testing a Radio Transmitter

Objective

To build a typical radio transmitter from a commercial kit and to perform tests for correct operation.

Materials

1   Ramsey Electronics (www.ramseykits.com) QRP20/30/40C transmitter kit and QAMP20/30/40C linear amplifier kit
    - Soldering iron and basic hand tools (side cutters, needle nose pliers, wire stripper)
    - Dummy loads (two 100-Ω 2-W resistors and MFJ-260C)
    - RG-58/U coax cable
    - RCA phono plugs
    - SO-239 UHF plugs
1   12–15-V power supply, 4 to 5 A
1   oscilloscope (25 MHz or greater vertical bandwidth)
1   RF power meter (Bird 43, MFJ-816 or equivalent)

Introduction

A commercial kit is one of the fastest, easiest, and lowest-cost ways to examine and test a contemporary HF transmitter. The one recommended here is designed for low-power (QRP) amateur radio use. In is available for three different operating frequencies in the ham bands, 20 (14 MHz), 30 (10 MHz), and 40 (7 MHz) meters (QRP20C, QRP30C, QRP40C). It does not make any difference which you choose for this project. Just keep in mind that if you do not have a valid FCC radio amateur license, you may not connect this transmitter to an antenna. Serious fines or even jail time can result from operation without a license. In this project, you will use a resistor as a dummy antenna load so no radiation will take place.

The transmitter is made up of a crystal oscillator, a buffer power amplifier, and a final amplifier that can deliver about 0.75 to 1 W to a 50-Ω load.

Measuring power accurately at low power is difficult, so it is strongly recommended that you add the linear power amplifier kit that produces up to 20 W of power into a 50-Ω load. The Ramsey models are QAMP20C, QAMP30C, and QAMP40C.

The circuit is a push–pull linear amplifier implemented with MOSFETs. The power from the transmitter is sufficient to drive the amplifier to full 20-W output. But be sure to get the kit matching the frequency of the transmitter.

To test the transmitter, you will need a dummy load that you can build with 100-Ω 2-W resistors. You should use a commercial dummy load for the power amplifier such as the MFJ-260C.

Procedure

1. Build the transmitter kit first following the instructions in the Ramsey manual.
2. Build the dummy load as illustrated in Fig. 8-3.
3. Build a shorting key as shown in Fig. 8-4. This transmitter is designed to be turned off and on by a telegraph key. The key is nothing more than just a convenient hand-operated switch. For this project, you can just use a shorted RCA phono plug. Just solder a short piece of hookup wire between the tip and shell of a phono plug. Plug the short into the KEY jack on the transmitter at the appropriate time.

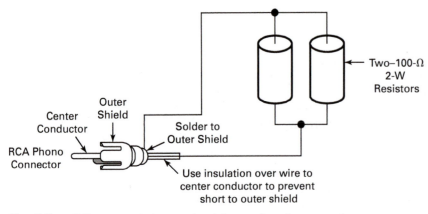

Fig. 8-3 Building a 50-Ω dummy load for testing the transmitter.

4. Connect the dummy load to the phono jack labeled ANT.
5. Connect the transmitter to the power supply. Be sure to set the voltage to 12 V if this is a variable supply before you turn the power on. The current drawn by this transmitter is about 400 mA with full output power.
6. Turn on the power and follow the instructions for testing in the Ramsey manual.
7. Plug the shorted phono plug into the KEY jack on the transmitter.
8. To verify the output power, connect an oscilloscope across the 50-Ω load and measure the voltage.
9. Remembering to use rms voltage, calculate the output power ($P = V^2/R$).
   $P =$ _____ watts
10. Describe the shape of the output signal.
11. Turn off the power. Remove the shorting plug.
12. Build the power amplifier kit following the instructions in the Ramsey manual.
13. Connect the power amplifier to the transmitter, dummy load, and power meter following the diagram in Fig. 8-5.
14. Connect the power amplifier and the transmitter to the power supply. Be sure to read the Ramsey manuals to ensure you understand the power supply requirements. Maximum power output of 20 W can be achieved with a DC supply voltage of about 13.8 V. The transmitter itself may operate at that voltage too and will provide about 1 W of drive to the linear amplifier. Do not exceed 15 V under any conditions.
15. Plug the shorting plug into the KEY jack on the transmitter.
16. Connect an oscilloscope across the 50-Ω load and measure the voltage.
17. Calculate the output power ($P = V^2/R$).
    $P =$ _____ watts
18. Describe the shape of the output signal. Explain.
19. Measure the output power on the MFJ-260C or other RF power meter. Follow the calibration and measurement procedures explained in the RF power meter manual.
20. Turn off the power. Remove the shorting plug.

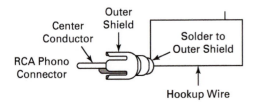

Fig. 8-4 Building a short to key the transmitter.

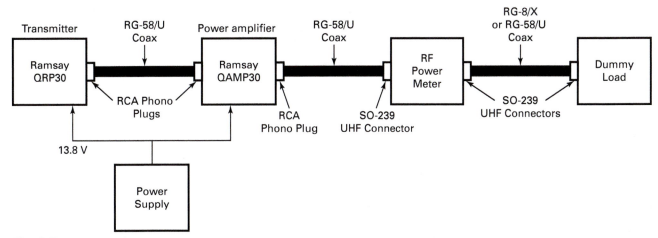

Fig. 8-5 Test setup for transmitter power measurement.

Questions

1. Referring to the schematic diagram of the transmitter in the Ramsey manual, state the type of crystal oscillator used.

2. Explain how it is possible to vary the frequency of this transmitter.

3. What class of operation is used by the buffer and final amplifiers in the transmitter?

4. The transmitter is designed to share a single antenna with a receiver. How is antenna switching implemented in this transmitter?

5. Referring to the schematic diagram of the power amplifier in the Ramsey manual, state the most likely operating class for this amplifier.

6. Describe the bias circuit for this amplifier.

7. What is the purpose of L1–L3 and C4–C7?

8. Is this amplifier resonant to any particular frequency? Explain.

9. Describes what happens when the RF signal from the transmitter is applied to J2. Explain the operation of the circuit with D2–D3 and Q3–Q5.

Project 8-4 Measuring Transmitter Output Power

Objective

To demonstrate how to measure transmitter output power and compute SWR.

Required

1   oscilloscope
1   RF power wattmeter (Bird model 43, MFJ-816, or equivalent, capable of forward and reverse power measurements, and of measuring 5 to 100 W up to 30 MHz)
1   radio transmitter portion of an amateur radio transceiver, capable of generating 1 to 100 W output in the 2- to 30-MHz range; alternatively, radio transmitter kits such as the Ramsey Electronics QRP 20/30/40 with QAMP 20/30/40 power amplifier
2   dummy loads, 50 Ω and 100 Ω, capable of dissipating power equal to the transmitter power
2   coax cables each about 2 or 3 ft long (RG-8/U or equivalent with 50-Ω impedance) with appropriate connectors such as N-type or PL-259

Introduction

The output impedance of a typical transmitter is usually 50 Ω. To achieve maximum power output, a 50-Ω antenna load is connected. If the antenna represents some other impedance, as is often the case, maximum power will not be sent to the load. What happens is that the power sent to the load is not completely absorbed by the load when the impedances are mismatched. This causes any unabsorbed power to be reflected back toward the transmitter. The forward power sent to the load and the reflected power add algebraically on the transmission line and produce what are called standing waves. Standing waves are patterns of voltage and current along the line.

The reflected power represents lost power. A measure of this lost power is the standing-wave ratio (SWR). You can easily determine the SWR through your power measurements. All you have to do is to measure the forward (P_F) and reflected (P_R) power and plug those values into the following formula to find SWR:

$$SWR = \frac{1 + \sqrt{P_R/P_F}}{1 - \sqrt{P_R/P_F}}$$

You can also compute SWR by finding the ratio of the load resistance (R_L) to the output impedance of the transmitter (Z_0):

$$SWR = \frac{Z_0}{R_L} \quad \text{or} \quad \frac{R_L}{Z_0} \quad \text{(SWR must be greater than 1)}$$

If the load is matched (equal) to the output impedance of the transmitter, the SWR is 1 and maximum power transfer is achieved. There is no reflected power. If the load impedance does not equal the transmitter output impedance, there will be some reflected power and the SWR will be greater than 1. Most transmitters can accommodate some mismatch, but few can accommodate an SWR of greater than 2, as damage may result from the reflected power.

In this experiment, you will learn to measure the output power of a transmitter. You will also measure the reflected power and compute SWR. Refer to Chapter 13 of the text for more details, or wait and run this experiment when you have completed Chapter 13.

Procedure

1. Read the operating manual for the transceiver or have your instructor explain it.
2. Set the transceiver to transmit in the CW mode. Set the frequency to a common ham band frequency, such as 14.05 or 10.16 MHz.
3. Refer to Fig. 8-6(a). Connect the 50-Ω dummy load to the transmitter output.
4. Turn on the power and tune the transmitter for maximum output as described in the transceiver manual.
5. Connect the scope across the dummy load. Measure the voltage across the load. (*Note:* Be very careful, as you could receive a shock or serious RF burn!)
6. Calculate the output power.
7. Turn off the power and connect the wattmeter into the line between the transmitter output and the dummy load as shown in Fig. 8-6(b).
8. If you have not already done so, familiarize yourself with the wattmeter by reading the instruction manual or asking your instructor. There should be provision for setting the power level to be measured and selecting a forward or reverse power measurement.
9. Turn on the power. Set the wattmeter to read forward power. Record the maximum output power you observe. How does this value compare to the value you computed in the step above? Explain.
10. Set the wattmeter to measure reflected power. Record the value of reflected power you measure.
11. Compute the standing wave ratio on the coax cable.
12. Turn off the power and disconnect the 50-Ω dummy load. In its place, connect the 100-Ω dummy load.
13. Turn on the power and measure the forward and reflected power output of the transmitter. Can you obtain the same power level as with the 50-Ω load? Explain your results.
14. Calculate the SWR.

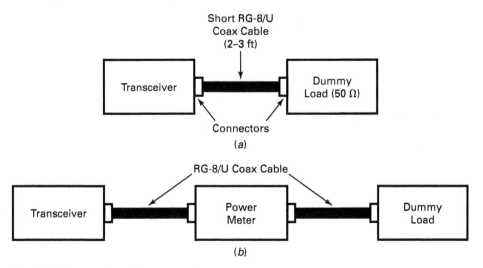

Fig. 8-6 Connections for measuring power.

Questions

1. What is the most desirable level of reflected power in a transmission line?
2. If the transmitter is designed for a 75-Ω output load and the antenna has an impedance of 20 Ω, what is the SWR?
3. The most desirable value of SWR is
   a. 0
   b. 1
   c. 2
   d. Infinity
4. Assume a forward output power of 150 W and a reflected power of 30 W. What is the SWR?
5. What is the maximum allowable SWR in most transmitters?
   a. 0
   b. 1
   c. 2
   d. Infinity

Project 8-5 Impedance Matching

Objective

To design, build, and test an *L* network to match a transmitter to an antenna.

Required

1   oscilloscope
1   function generator (50-Ω output impedance and capable of a frequency of 2 MHz)
1   51-Ω resistor
1   22-Ω resistor
1   each, 0.0018-, 0.002-, 0.0022-, and 0.0027-μF capacitors
1   each, 1.8-, 2.2-, and 4.7-μH inductors

Introduction

One of the most important parts of any transmitter is the output impedance-matching network. Various *L, T,* and pi (π) networks are used between transmitter stages and between the final amplifier and the antenna to ensure maximum power transfer. As a technician, you will not likely be involved with interstage impedance-matching networks, but you will often encounter situations where it is necessary to create a special impedance-matching network between a transmitter and an antenna. An *L* network is frequently used to match the transmitter to the antenna impedance. This provides minimum SWR and maximum power transfer to the antenna. Commercial impedance-matching units are available for this purpose. They are generally referred to as antenna tuners and contain a variety of variable capacitors and inductors that can be connected and adjusted in a variety of ways to achieve an impedance match.

Although antennas are designed to be resistive loads, in most cases they are not. If they are operated at a frequency even slightly different from the design center frequency, their impedance will be reactive. Furthermore, the overall antenna impedance depends on the height above the ground and the effects of any nearby structures and objects. To improve the impedance match between transmitter and antenna, an *L* network is introduced between the transmitter output and the coax cable to the antenna.

A commonly used *L* network is shown in Fig. 8-7(*a*). This variation of the *L* network is used when the load impedance is less than the output impedance of the transmitter, which is a common occurrence. It is also a low-pass filter, so it provides some attenuation of unwanted harmonics that almost always appear in the output of a transmitter. Values of *L* and *C* can be computed to make the antenna impedance appear to be equal to the transmitter output impedance.

If you redraw the circuit as shown in Fig. 8-7(*b*), you can see how this is done. The coil and capacitor actually form a parallel resonant circuit where the resistive antenna appears in series with the coil and becomes the coil's apparent series resistance. The goal of the calculations is to select values of *L* and *C* such that the parallel circuit looks resistive at the frequency of operation and so that its impedance is equal to the transmitter output impedance.

Recall that you can find the impedance of a parallel resonant circuit (Z_p) with either of the following formulas:

$$Z_p = \frac{L}{CR_s}$$

$$Z_p = R_s(Q^2 + 1)$$

Questions

1. The frequency variation increment of a PLL synthesizer is determined by the _____ .
   a. VCO free-running frequency
   b. Frequency divider ratio
   c. Both **a** and **b**
   d. Reference frequency

2. A PLL synthesizer has a 1-MHz reference and a frequency divider of three cascaded BCD counters. The output frequency is _____ .
   a. 1 MHz
   b. 10 MHz
   c. 100 MHz
   d. 1 GHz

3. In the circuit described in question 2, a two-decade BCD counter is connected between the reference and the PLL phase detector input. The output frequency is _____ .
   a. 10 kHz
   b. 100 kHz
   c. 10 MHz
   d. 100 MHz

4. (True or False) The PLL synthesizer can function as a frequency multiplier.

5. To function properly, the VCO in the PLL must have its free-running frequency set approximately to the _____ .
   a. Desired output frequency
   b. Output divided by N
   c. Frequency division ratio
   d. Reference frequency

Project 9-3 Mixer Operation

Objective

To demonstrate mixer operation using a diode ring, lattice mixer.

Required

1 crystal oscillator circuit from Project 8-2
1 signal generator with an output frequency range from 11 to 13 MHz
   (MFJ-269 or comparable)
1 frequency counter
1 Mini-Circuits SBL-1-1 + frequency mixer
1 Toko 455-kHz ceramic filter from Project 2-3
2 2-kΩ resistors
1 0.001-μF ceramic capacitor

Introduction

There are literally dozens of mixer configurations used in modern communications equipment. From a simple diode to a complex Gilbert cell IC, all have
their place. One of the best and most widely used mixers is the diode ring or
lattice modulator. While it does introduce conversion loss, its performance is
superior to most other types of mixers. This type of mixer is a balanced
modulator type, meaning that it suppresses the carrier or, in this application,
the local oscillator signal in the output. This leaves only the sum and difference frequencies in the output. The desired frequency is then selected by a
filter. In this project, you will show the operation of such a mixer and how
the desired output is selected with a ceramic filter.

Procedure

1. Construct the circuit shown in Fig. 9-6. Be sure to wire the mixer as shown
    in Fig. 9-7. The circuit uses the crystal oscillator circuit from Project 8-2
    that produces a 12-MHz output. This will be used as the local oscillator
    (LO). An external signal generator will be used for the signal source. This
    generator should be variable so that its frequency can be varied. A
    frequency counter is needed to measure the signal frequency from the
    generator. If the MFJ-269 is used, the counter is internal and a separate
    unit is not needed. The mixer is a commercial diode ring-type mixer
    designated as the Mini-Circuits SBL-1-1+. The 455-kHz band-pass
    ceramic filter used in Project 2-3 will serve to extract the sum or difference
    frequency.
2. Looking at the circuit, determine what the input signal frequency from
    the generator should be to produce a difference intermediate frequency
    (IF) of 455 kHz. There should be two frequencies. Also compute the related
    sum frequencies of this mixing result.
3. Power up the circuit and increase the signal generator output to its max-
    imum. While monitoring the filter output with an oscilloscope across the
    2-kΩ load, tune the signal generator from about 10 MHz upward until
    you see a peak output voltage. Note the frequency. Continue to increase
    the generator frequency while looking for a second input frequency that
    produces a peak output voltage. Again note that frequency.
4. Do the frequencies you measured in step 2 correspond to those you mea-
    sured in step 3? Explain. Do the sum frequencies appear in the output?

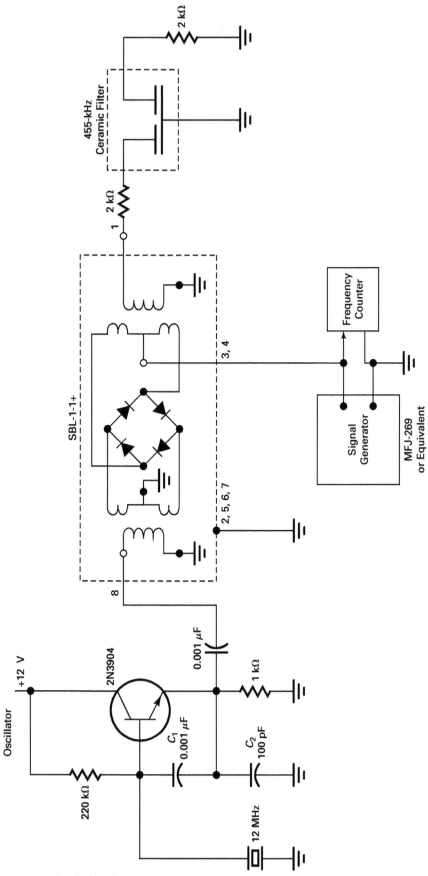

Fig. 9-6 Circuit for demonstrating mixer operation.

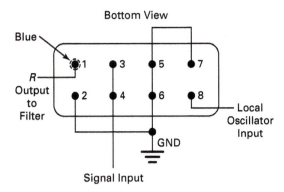

Fig. 9-7 Wiring for the SBL-1-1+.

Project 9-4 Building and Testing a Communications Receiver

Objective

Gain hands-on experience with actual communications components, equipment, and testing procedures by building a commercial kit receiver.

Required

1 Ramsey Electronics Inc. (www.ramseykits.com) FR146 FM receiver kit
1 MFJ Enterprises Inc. (www.mfjenterprises.com) MFJ-269
  HF/VHF/UHF SWR analyzer
• Soldering iron (<35 W), solder, hand tools.
• Hookup wire

Introduction

There is probably no better way to understand and appreciate the complexity of modern communications equipment than by actually working with it. One way to do this inexpensively and in a relatively short period of time is to build a kit. A kit is a collection of components that, when assembled and tested, forms a complete workable end product. There are a variety of electronic kits that can be quickly and easily assembled with a soldering iron and some basic hand tools. The end product is a workable electronic product. Many radio receiver kits are available, from simple AM and FM radios to more sophisticated amateur radio transceivers. This project recommends a specific communications receiver, the Ramsey FR146. It is a dual conversion superheterodyne receiver for receiving VHF FM radio signals. It is targeted at the amateur radio market because it covers the ham radio 2-m band that extends from 144 to 148 MHz. The receiver can actually be tuned to any 5-MHz segment in the 136- to 175-MHz range. This wide range includes most two-way FM radio communications in the United States and elsewhere. Marine, government, public service, and public safety (police, fire, etc.) and other services use segments in the range. Therefore this receiver is very typical of many in use today and provides a good look at the circuits and techniques used.

In this project, you will build the kit, analyze the receiver circuit, examine some of the integrated circuits in detail, test and tune the receiver, and use it.

Procedure

1. Do an analysis of the circuits before you build the receiver. Read the introduction and circuit description of the FR146 receiver on pages 4 and 5 in the Ramsey instruction manual. Be sure to examine the block diagram on page 8 and the schematic diagram on pages 12 and 13.
2. As part of the circuit analysis above, it will be helpful if you will find and print out the data sheets for the integrated circuits used in this receiver. You can find these data sheets on the IC manufacturer's websites:
   NE602   National Semiconductor (now Texas Instruments)
   MC3359   Freescale Semiconductor
   LM386   National Semiconductor (now Texas Instruments)
   Reading the data sheets will help you in the circuit analysis.
3. Once you have read the circuit description, answer the questions in the section at the end of this project. Build the FR146 kit, following the instructions in the Ramsey manual.
4. Once you have built the kit, test it using the procedure described. This project recommends the following variations to the procedures on pages 18 and 19 in the Ramsey manual.

a. Use the MFJ-269 as a signal source to test the receiver.

b. If the MFJ-269 is not available, an excellent alternative source is a nearby NOAA weather station.

c. Go to the National Oceanic and Atmospheric Administration website and to their National Weather Service (NWS) page, www.nws. noaa.gov/nwr/nwrbro.htm. Find the nearest weather radio station near you. These FM stations are in the 162-MHz range and are perfect for testing your radio. Find the frequency for your station so you can set the local oscillator frequency in the receiver as you align it.

d. If you set the receiver to the approximate center of the NOAA/NWS band, tune for other stations in your area. You may need a better antenna. Use a wire at least 18 inches long for the antenna.

Questions

1. The FR146 is a dual conversion superheterodyne. What are the two IFs?
2. What is the main function of the NE602 IC?
3. What is the function of L3 and D1?
4. What type of tuned circuit is used for IF filtering?
5. What is L4 used for? What circuit is it associated with?
6. Where is the second IF mixer located?
7. What is the second local oscillator frequency, and how is it set?
8. Does this receiver have AGC? Squelch? Where are they located, and what components are involved with each?
9. Explain the circuit using Q2. What is its function? What is wrong with it?
10. If the DC supply voltage is 9 V, where does the 6-V supply line come from?

Chapter 10

Multiplexing and Demultiplexing

Project 10-1 Pulse Amplitude Modulation and Time Division Multiplexing

Objective

To test a circuit that will permit demonstration of both pulse amplitude modulation and time division multiplexing.

Required

1   oscilloscope
1   sine-wave signal source
1   square-wave signal source
1   555 timer IC
1   4016 or 4066 quad bilateral CMOS switch IC
1   NPN transistor (2N3904, 2N4401, etc.)
2   10-kΩ pots
1   4.7-kΩ resistor
2   15-kΩ resistors
1   22-kΩ resistor
1   0.01-μF capacitor
1   10-μF electrolytic capacitor

Introduction

In pulse amplitude modulation, the information signal to be transmitted is turned off and on periodically by a gate circuit. The gate, usually some kind of switch, allows segments of the intelligence signal to be passed through. The process is generally known as sampling the intelligence signal. The output of the gate is a series of periodic pulses whose amplitude follows the intelligence signal. This series of pulses is then used to modulate a carrier.

A key design factor in a PAM modulator circuit is the rate of sampling. To represent the intelligence signal accurately, the sampling rate must be at least 2 times the highest frequency content contained in the modulating signal. Higher sampling rates of 10 or more samples for the highest frequency in the intelligence signal are more desirable because they more accurately represent the information being transmitted.

The concept of PAM can be expanded to produce time division multiplexing. A time division multiplexer has two or more inputs or channels, each of which is sampled at a high rate of speed. A gate on each input intelligence signal samples its respective signal for a short period of time. Each of the signals is sampled one at a time, in sequence so that a segment of each signal appears in the output. For example, a four-input time division multiplexer would sample four input signals sequentially, one after the other. The cycling sequence then repeats again and again. The resulting time-division multiplexed output signal is a composite of the four intelligence signals sampled sequentially. The result is that four intelligence signals can be transmitted simultaneously over a single channel.

In this experiment, you are going to demonstrate PAM and time division multiplexing. You will use a solid-state switch that is operated by an astable multivibrator. You will observe both PAM and time division signals at the output.

The gate or switch used in this experiment is a CMOS switch. It is made of both P- and N-channel enhancement-mode MOSFETs connected in such a way that they act as a simple single-pole, single-throw switch. When a binary 1 logic signal is applied to the switch, the switch turns on. When a logic signal is binary 0, the switch turns off. During the on period, the intelligence signal is passed. The 4016 and 4066 ICs contain four sampling switches.

One of these switches will be used to demonstrate PAM, and two of them will be connected together to form a two-channel multiplexer. A 555 IC timer connected as an astable multivibrator provides the sampling signal.

Procedure

1. Construct the circuit shown in Fig. 10-1. The 555 IC is used as an astable multivibrator to generate a square wave that will operate the CMOS

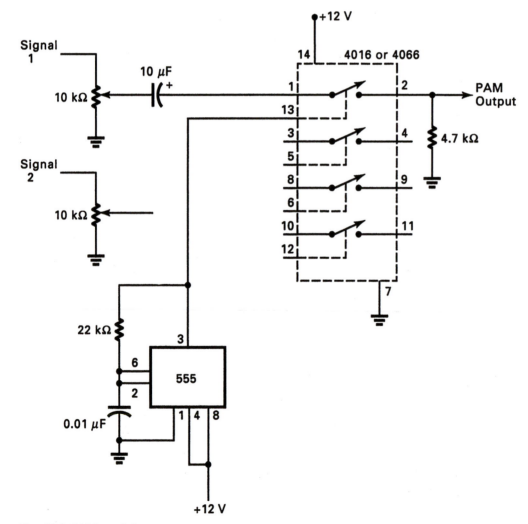

Fig. 10-1 PAM modulator.

switch. The resistor and capacitor associated with the 555 sets the 555's oscillating frequency. The 4016 or 4066 IC contains four CMOS switches, and you will use one of them as a gate to produce a PAM modulator circuit. Your signal source can be a function generator or any other source of a sine-wave intelligence signal. Apply the sine wave to the signal 1 input. The 10-kΩ pot will be used to adjust the signal amplitude. The output is taken from across the 4.8-kΩ resistor at the switch output. Check to be sure that each integrated circuit is connected to +12 V and ground before beginning.

2. Set the function generator supplying signal 1 for 150 Hz with an amplitude of several volts peak to peak. Then while monitoring the signal at the arm of the 10-kΩ potentiometer, adjust the pot for a signal amplitude of approximately 0.5 V_{p-p}.

3. Apply power to the circuit. With an oscilloscope, observe the 555 oscillator output at pin 3. Use the calibrated horizontal sweep on the oscilloscope or a frequency counter to determine the frequency of the sampling signal. Record your result.

   Considering the frequency of the input intelligence frequency and the sampling frequency above, is the intelligence signal sampled fast enough to retain intelligibility?

4. Connect the oscilloscope to the 4.7-kΩ output resistor. You should observe a PAM output signal. Each time the sampling signal is a binary 1, or approximately ±12 V, the switch closes. That allows a portion of the input signal to appear at the output. When the sampling oscillator produces a binary 0 output, the switch is open, thereby causing the output to be zero. In observing the PAM output, you should note that the switch passes both the positive and negative alternations of the input sine wave.

5. Modify the modulator circuit as shown in Fig. 10-2. You are connecting a pair of 15-kΩ resistors as a dc voltage divider to the gate input, and that provides a dc offset for the input signal. In other words, the ac input signal will be riding on a dc signal, as will the output.

6. Observe the output waveform. Note that the sampled output sine wave is riding on a dc level. The signal is offset from the zero baseline by a voltage level equal to that supplied by the voltage divider. The dc offset at the input to the switch is approximately equal to how many volts?

7. Turn off the power and modify the circuit so that it appears as shown in Fig. 10-3. Remove the two 15-kΩ voltage divider resistors, and then connect the second 10-kΩ pot to another of the solid-state switches in the 4016 or 4066 IC. Connect the switch output to the 4.7-kΩ resistor output as shown. Next, wire the transistor inverter circuit. This switching inverter provides a complement signal to the second gate switch. When the upper switch is on, the lower switch will be off, and vice versa. That means the two input channels will be alternately sampled depending on which switch the binary 1 signal is applied to.

8. Apply power to the circuit. Initially do not connect an input at the signal 2 pot. Set the pot for zero input signal and observe the output. At this time, you should see a signal like that you obtained with pulse amplitude modulation. You are still observing a sampled version of the 150-Hz sine wave applied to signal input 1. However, during the time that the upper (A) switch is open, the lower (B) switch is closed. The signal at the No. 2 input is 0 V; therefore, you are seeing a 0-V level during the time that the B switch is closed. Applying another signal to the No. 2 input will cause that signal to appear at the output during the time the B switch is closed.

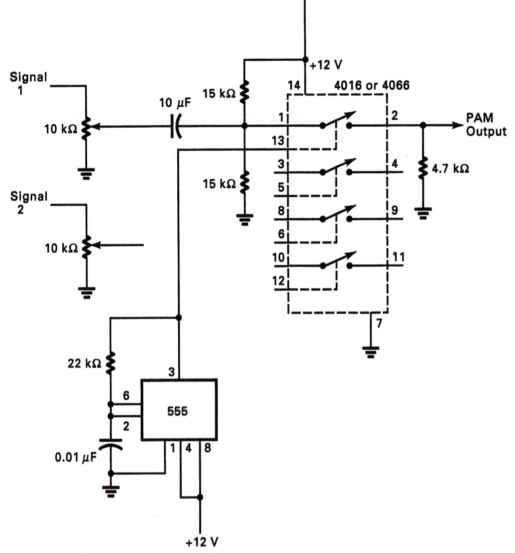

Fig. 10-2 PAM modulator with dc offset.

9. Connect a positive dc voltage to the signal 2 input. Adjust the dc voltage at the arm of the 10-kΩ pot for a value of approximately 0.25 to 0.3 V. Now observe the multiplexed output. What you are observing is the sampled 150-Hz sine wave and a dc voltage level other than zero. While observing the output signal, adjust the 10-kΩ pot on input signal 2 so you can see the output variation. You can take this one step further and reverse the polarity of input signal 2 to show the effect.

10. Remove the dc voltage from the signal 2 input. Using a second function generator, connect a square wave to the signal 2 input. Set the frequency for approximately 200 Hz and an output that varies between zero and +5 V.

11. Observe the multiplexed output waveform. With all of the different signals involved—the 150-Hz sine wave, 200-Hz square wave, and the sampling oscillator signal—your oscilloscope will go crazy. By using the horizontal frequency and the triggering controls, you should be able to obtain a stable output waveform. Making minor adjustments in the square-wave input frequency will also help to stabilize the waveform for viewing. Of course, what you should see is a sine wave and a square wave appearing in the output, but sampled. The samples are, of course, interlaced, which causes the signals to be merged. If additional input signals were connected,

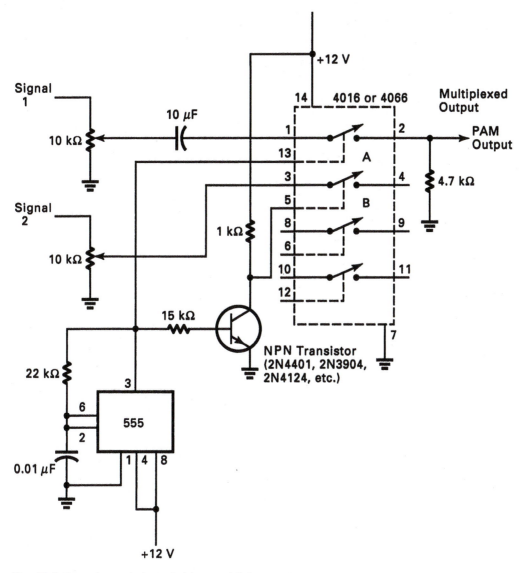

Fig. 10-3 Two-channel time division multiplexer.

these two would be merged with the composite signal. The resulting output signal is relatively complex and meaningless by itself. However, when it is applied to a demultiplexer circuit, the different signals will be unscrambled and the original modulating signals will be recovered intact.

Chapter 11

Digital Data Transmission

Project 11-1 Serial Digital Data

Objective

To demonstrate one way to develop serial binary data and how to interpret an oscilloscope display of that data.

Required

1 dual-trace oscilloscope
2 4035B 4-bit CMOS shift register ICs
1 555 timer IC
1 10-bit DIP switch
11 4.7-kΩ resistors
1 22-kΩ resistor
1 0.01-μF capacitor
1 100-μF electrolytic capacitor

Introduction

In virtually all data communications applications, binary data is transmitted serially. That is, rather than transmit all bits of a binary word simultaneously or in parallel, the data is transmitted one bit at a time. The reason for this in communications applications is obvious. To transmit parallel data, multiple paths or channels, one per bit, are required. In most communications systems, only a single channel is available. For that reason, multiple binary words are transmitted one after another, one bit at a time.

On the other hand, in electronic equipment that is not used for communications, binary data is usually generated and manipulated by parallel methods. This is particularly true of computers. For this reason, the processes of parallel-to-serial and serial-to-parallel data conversions are important.

The circuit that is most commonly used to perform parallel-to-serial and serial-to-parallel data conversions is the shift register. A shift register is a digital circuit made up by cascading D or JK flip-flops. To translate a parallel binary number into a serial number, the parallel number is loaded into the shift register flip-flops, and then a clock signal is applied simultaneously to all of the flip-flops. The resulting data is shifted one bit to the right (or to the left in some cases), thereby generating a serial train of data at the output flip-flop.

The process of parallel-to-serial conversion is illustrated in Fig. 11-1. An 8-bit parallel binary number is loaded into the flip-flops of an 8-bit shift register, and then clock pulses are applied to all flip-flops simultaneously. When the first clock pulse occurs, all of the bits in the binary number are shifted one position to the

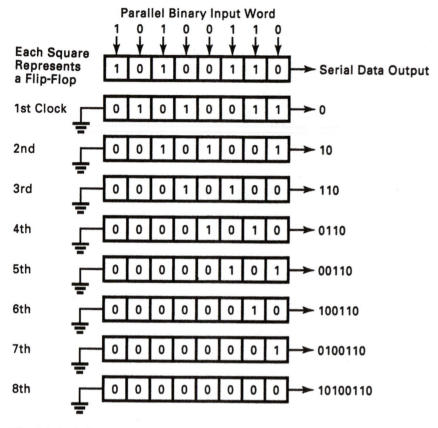

Fig. 11-1 Shift register used for parallel-to-serial data conversion.

right. The bit in the right-hand flip-flop is shifted out. That bit is then applied to some other circuit or transmitted over the communications channel. An example is the modulator in a modem for transmission over telephone lines.

When the second clock pulse occurs, all of the bits are again shifted one position to the right. The bit in the rightmost position is shifted out. With each succeeding clock pulse, the bits in the shift register are moved to the right as the number is shifted out. By the time the eighth clock pulse occurs, all bits have been shifted. Note that the input to the left-hand flip-flop of the shift register is connected to ground. That represents a binary 0. As the data is shifted to the right, binary 0s are shifted into the flip-flops.

If you were to observe the serial data output of the shift register of Fig. 11-1, you would see a signal like that shown in Fig. 11-2. The upper waveform

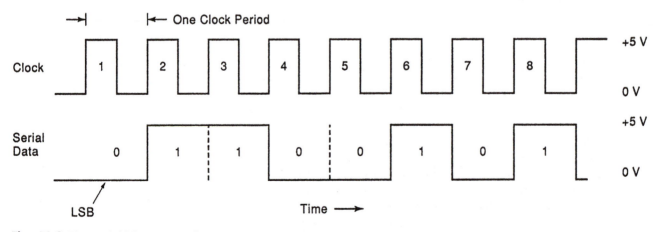

Fig. 11-2 The serial binary waveform.

106

is the clock, a series of periodic pulses. The clock pulses are numbered to correspond to the clock signals designated in Fig. 11-1.

The serial data shows the voltage waveform at the output of the right-hand flip-flop in the shift register. After the first positive-going transition of clock pulse 1, a binary 0 output appears. When the second positive-going transition of clock pulse 2 occurs, a binary 1 appears at the output. Note the sequence of the bits in Fig. 11-2 and compare it to the sequence of the bits in Fig. 11-1. They appear to be reversed. The reason is that an oscilloscope displays voltage levels with respect to time, with time proceeding from left to right. You will see the binary data displayed from left to right on the screen as it is shifted out.

An important consideration in transmitting serial data is which bit of a binary word is transmitted first, the least significant bit (LSB) or the most significant bit (MSB). This is important to know so that you can interpret the data after it has been transmitted. Various schemes are used in data communications systems, in which either the MSB or LSB is transmitted first. You must become familiar with the protocols and the system specifications to determine this information.

For example, when ASCII data is transmitted, the LSB is usually transmitted first. Recall that the ASCII system represents letters of the alphabet, numbers, punctuation symbols, and other characters as 7-bit binary numbers. The serial word in Fig. 11-2 might represent the ASCII 0100110. The LSB in that case would be the zero at the left-hand side of the waveform. The eighth bit, which is a binary 1, will usually represent a parity bit. The ASCII number in Fig. 11-2, 0100110, therefore represents the ampersand (&). The eighth or parity bit being a binary 1 means that an even-parity system is used because there are an even number (4) of binary 1s in the total transmitted value. Note in Fig. 11-2 that, since the LSB is transmitted first, it appears on the left because it occurs earlier in time than the other bits.

In this experiment, you are going to demonstrate the generation of serial binary data by using a shift register. Two 4-bit CMOS shift registers are combined to form an 8-bit shift register. Parallel data will be loaded into this register by way of a set of DIP switches. An oscillator made up of a 555 timer IC will be used to produce the clock pulses to cause the data to be shifted out. In order that you can view the data on an oscilloscope, the serial data output from the shift register is connected back to the serial input of the shift register. In other words, as the data is shifted out, it is also shifted back into the same register. That allows the word to be shifted out and then reloaded again and again for repetitive viewing on an oscilloscope. By loading in different binary values and displaying the clock pulses and serial data output simultaneously, you will be able to determine the relationship between the parallel input bits and the serial output bits.

Procedure

1. Construct the circuit shown in Fig. 11-3. Because of the complexity of the circuit, be extremely careful in making the interconnections to avoid wiring errors.

   The pin-out diagram for the 4035B shift register is shown in Fig. 11-4; that is how the pins on the 16-pin DIP are numbered when you view the IC from the top. The P1 through P4 pins accept a 4-bit parallel binary input word. The individual flip-flop outputs are designated Q1 through Q4; Q1 is the flip-flop nearest the serial input. The serial input is applied to the J and K NOT inputs on pins 3 and 4. Usually these two pins are tied together. Pin 6 accepts the CLOCK input. Pin 5 is the reset (RST) line. Bringing this line high or to binary 1 causes all flip-flops to be simultaneously reset. The LOAD input at pin 7 is used to cause the parallel binary

the clock waveform on the upper trace of the oscilloscope. One complete 8-bit word of serial data will then be displayed on the lower trace.

5. To see the variation in the output waveform based on the actual value loaded into the shift register, try loading in different binary numbers and observing the binary output. The following are some examples:

10101010

00110011

11101110

00001111

In each case, set the DIP switches to apply the number to the shift register. The LSB shown above will be the value loaded into what we are calling the LSB of the shift register, which in this case is input pin 9 on the No. 1 4035B IC. The MSB then is pin 12 on the No. 2 4035B IC. Once the DIP switches have been set to the desired number, momentarily apply a binary 1 to pin 7 of the ICs with the load DIP switch. You should immediately note a change in the output pattern on the oscilloscope screen. Be sure that you can relate the binary numbers above to the corresponding oscilloscope patterns that they create.

6. Based on the clock frequency measured in the first step, how long does it take to transmit a single 8-bit binary word?

7. Draw a diagram similar to the one in Fig. 11-1 that illustrates how a serial binary data word is captured by shifting it into a shift register. Illustrate the process in the same number of steps to show how serial-to-parallel data transfer occurs. Use the ASCII code for the letter M with odd parity.

Project 11-2 Frequency Shift Keying

Objective

To demonstrate the generation of a frequency shift keying signal.

Required

1 dual-trace oscilloscope
1 square wave signal source
1 2206 function generator IC
1 150-kΩ resistor
2 4.7-kΩ resistors
1 22-kΩ resistor
1 47-kΩ resistor
1 56-kΩ resistor
1 0.001-μF capacitor
1 1-μF capacitor
1 10-μF capacitor

Introduction

Frequency shift keying is a modulation technique used to transmit binary data over analog communications channels such as the telephone lines. The binary data operates an FSK modulator that translates the binary signals into two different tones or sine waves. The frequencies chosen depend on the bandwidth of the analog channel as well as the speed of the binary data. Because telephone lines have a very narrow bandwidth (approximately 3 kHz), low frequencies are used. For example, in low-frequency modems, the frequencies of 1070 and 1270 Hz are used to represent the binary values 0 and 1, respectively. In full-duplex systems, the tones 2025 and 2225 Hz are also used to represent binary 0 and 1, respectively. When binary 0 occurs, the low-frequency tone is generated; when a binary 1 occurs, the high-frequency tone is generated. Other schemes have also been used. Because such low frequencies are used, the data rate is restricted; the maximum with such a system is typically in the 300 bits/s region. For higher data rates, other types of modulation such as PSK and QAM are used.

In this experiment you will demonstrate the generation of FSK and observe the output signal. You will use the 2206 function generator IC that you used in preceding experiments. This IC was designed specifically for FSK applications. The serial binary input will be simulated by a square wave.

Procedure

1. Construct the circuit shown in Fig. 11-5. You may have retained the 2206 function generator circuit used in many preceding experiments. If so, modify it so that it appears as shown in Fig. 11-5.
2. Apply a square wave to pin 9 of the 2206 IC; it will simulate a serial binary input. The square wave should have an amplitude that switches between 0 and approximately +15 V. Any TTL data source will be suitable. Set the frequency to approximately 1.5 kHz.
3. Apply power to the 2206. Monitor the FSK output at pin 2. Because the input signal repeatedly switches the output between two different sine-wave frequencies, scope triggering will be somewhat erratic. However, through a combination of scope triggering and horizontal sweep frequency control adjustments, you should be able to stabilize the waveform and see the two different frequencies as the input signal switches between binary

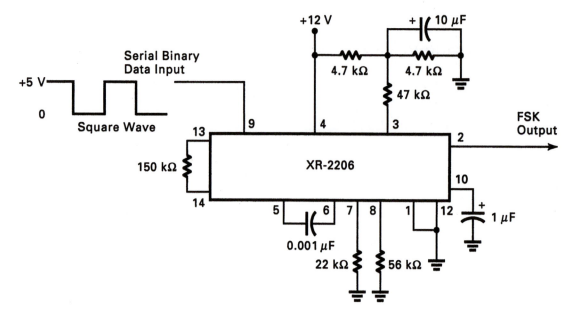

Fig. 11-5 FSK generator.

0 and binary 1. Use the variable sweep frequency and trigger level controls to stabilize the waveform. If you have a dual-trace oscilloscope, you can display the binary input signal on the upper trace and the FSK output signal on the lower trace.

4. Determine which binary input state produces which output frequency. One way to do that is to remove the square wave from pin 9 and then ground pin 9 to provide a fixed binary 0 input. Measure the 2206 output frequency at pin 2 with a frequency counter or on the oscilloscope. Apply +5 V dc to pin 9 and again measure that 2206 output at pin 2. That will simulate a binary 1 input. Record the information you have collected.

5. The frequencies of the output tones produced by the 2206 are determined by resistor and capacitor values; specifically, the 0.001-μF capacitor connected between pins 5 and 6 and the 22- and 56-kΩ resistors connected to pins 7 and 8 in Fig. 11-5. Knowing that, what steps would you take to reverse the frequency tones you determined in step 4? That is, how would you change the tone obtained with binary 0 so that it is produced when a binary 1 occurs, and vice versa?

6. Suggest a type of circuit that might be used to demodulate an FSK signal.

Project 11-3 Binary Phase-Shift Keying (BPSK)

Objective

To demonstrate a method of generating and detecting a BPSK signal.

Required

1   oscilloscope (dual-trace, 50-MHz minimum)
2   function generators (sine wave and TTL output)
1   frequency counter
2   Mini-Circuits SBL-1-1+ diode ring lattice modulators/mixers
1   Maxim MAX202CPE RS-232 line driver IC
1   74LS13 TTL IC
1   0.001-μF ceramic disk capacitor
6   0.1-μF ceramic disk capacitors
1   0.15-μF electrolytic capacitor
1   330-Ω resistor

Introduction

Binary phase-shift keying (BPSK) is one of the most widely used types of modulation in transmitting digital data. The digital data is used to shift the phase of a constant-frequency carrier. In BPSK, a binary 0 is transmitted as a carrier with no phase shift, while a binary 1 is transmitted with a 180° phase shift (phase inversion). This is illustrated in Fig. 11-6.

There are many variations of PSK. For example, quadrature PSK (QPSK) has four phase shifts spaced by 90°; 8PSK has eight phase-shift values, and so on. These forms of PSK permit each phase to represent multiple bits so that higher data can be achieved. With QPSK, each phase-shift value represents 2 bits of data. With QPSK, the data rate is twice what it is with BPSK.

In this experiment you will see a way to generate BPSK with a diode lattice modulator. The basic circuit is shown in Fig. 11-7. The sine-wave carrier is fed to a transformer, which applies the carrier to a set of diode switches. The diode switches determine the connection of the signal from the secondary

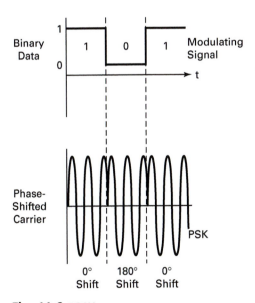

Fig. 11-6 BPSK.

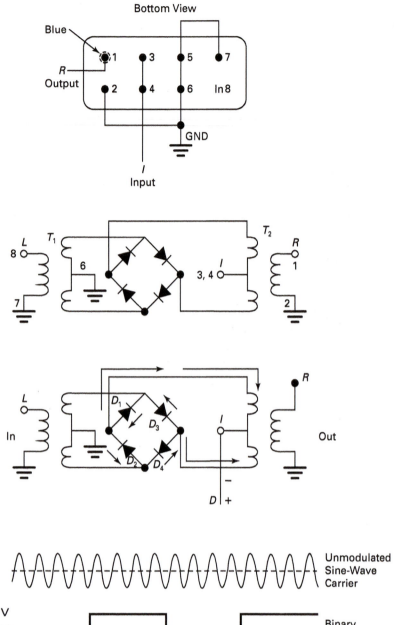

Fig. 11-7 Diode lattice modulator.

winding of T_1 to the primary winding of T_2. The output is taken from the secondary winding or T_2.

The state of the diodes (off or on) is determined by the binary input signal. This signal must be bipolar. A typical signal is $+5$ V to $+10$ V for a binary 1 to -5 V to -10 V for a binary 0, or vice versa.

Although this circuit can be constructed with discrete components, diode lattice modulators are available as a packaged device. One of these is the Mini-Circuits SLB-1-1+, which is also widely used as a mixer. These low-cost circuits have matched transformers and diodes. This circuit is of the form shown in Fig. 11-7. When a binary 1 is applied ($+5$ V), diodes D_1 and D_4 are forward biased by the input signal and conduct. D_2 and D_3 are cut off. The conducting diodes act like a short circuit and connect the signal from the secondary of T_1 to the primary of T_2. The windings are arranged so that no phase inversion occurs.

When a binary 0 occurs, the signal applied to the modulator is -5 V. This reverse-biases D_1 and D_4, causing them to cut off. D_2 and D_3 conduct and act as a short circuit to connect the upper lead of the T_1 secondary to the lower lead of the primary of T_2 and the lower lead of the secondary of the T_1 to the upper lead of the primary of T_2. This connection causes T_2 to produce a $180°$ phase shift at the output.

This experiment uses a special IC (Maxim MAX202CPE) to convert the conventional TTL or CMOS binary voltages into the ±10-V signals necessary for the proper operation of the diode lattice modulator. The circuit uses the $+5$-V supply voltage to power a circuit that automatically produces $+10$-V and -10-V sources to power the circuit. No external negative power supply is needed.

Procedure

1. Connect the circuit in Fig. 11-8. Note the bottom view of the SBL-1-1+ in Fig. 11-7, which shows the pin numbers that correspond to the numbers and connections shown in the schematic. Were the SBL-1-1+ first as shown

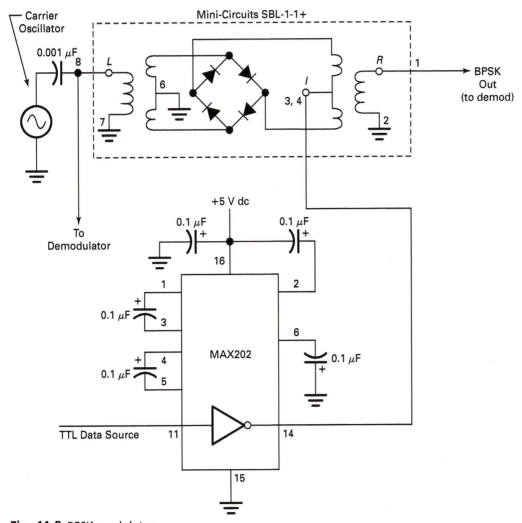

Fig. 11-8 BPSK modulator.

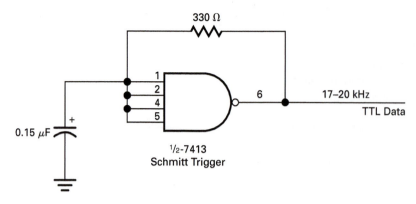

Fig. 11-9 A TTL Schmitt trigger used as a binary data source.

in Fig. 11-7, then connect it as shown in Fig. 11-8. You will use a function generator to produce the sine-wave carrier. The binary signal can be simulated by any standard digital source. If a second tunable digital function generator is not available, use the simple circuit shown in Fig. 11-9.

2. Turn on the function generator and set the sine-wave carrier level in the primary of the SBL-1-1+ to 0.5 V_{p-p} at a frequency of 300 kHz. Apply dc power to the remainder of the circuit. If you are using a separate function generator for the digital signal, set its frequency to 20 kHz. Use the TTL output from the function generator. If you build the circuit in Fig. 11-9, its frequency is already set to approximately 17 kHz if you used the *RC* values specified.

3. Now, observe the output of the MAX202 IC at pin 14 that is applied to pin 4 of the SBL-1-1+. Measure its voltage levels. Is it bipolar?

4. Next, observe the BPSK output at pin 6 of the SBL-1-1+. Use the second input channel of the scope and display the modulated output along with the modulating signal. Draw the waveforms that you see.

5. Adjust the frequency of the carrier generator until you see a BPSK waveform that is as close to those in Fig. 11-6 as possible. You will need to tune the function generator carefully to synchronize the carrier and modulating signal frequencies. Measure both the carrier and data frequencies with the counter.

6. Next, build the demodulator circuit shown in Fig. 11-10. The BPSK signal from the modulator is applied to the primary of T_1 and the carrier

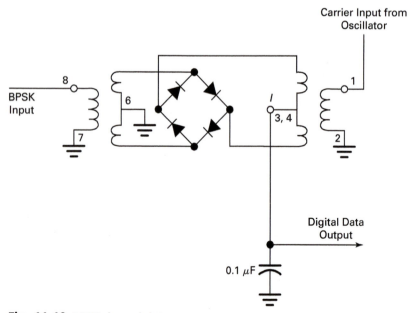

Fig. 11-10 BPSK demodulator.

116

is applied to pin 1 of the SBL-1-1+ in the demodulator. The output is taken from pin 4 of the SBL-1-1+ in the demodulator.

7. Observe the output of the demodulator. What do you see? Try adjusting the carrier frequency slightly to maximize the output signal. Measure the output voltage. Explain what you see.

Questions

1. Explain why the BPSK output does not start and stop perfectly at the zero crossing of the carrier sine wave. How does adjusting the carrier frequency help correct this condition?
2. Explain how the demodulator works.
3. In this experiment, you used the same carrier signal source for both modulation and demodulation. In separate transmitters and receivers, name the source of the receiver carrier signal and explain how it is derived.
4. What is the capacitor in the demodulator used for?

Project 11-4 Spread Spectrum

Objective

To further explore the technology of spread spectrum and its current usage.

Required

PC with Internet access.

Introduction

Spread spectrum is one of the most widely used radio technologies today, primarily thanks to its adoption by the cell phone industry as its primary method. However, it is also used in other applications. This project asks you to dig deeper into the spread spectrum method and discover how it works and where it is used.

Procedure

1. Go to the following websites and explore the information on spread spectrum. Take any interesting tutorials and print out other information such as white papers.

   www.wikipedia.com
   www.sss-mag.com
   www.palowireless.com
   www.commsdesign.com

2. Perform Internet searches on spread spectrum and related topics and examine the various documents produced.
3. Answer the questions below, using the resources you discovered.

Questions

1. What is the name given to spread spectrum as it is used in cell phones?
2. How is multiplexing of multiple users in cell phone systems accomplished with spread spectrum?
3. What company is largely responsible for developing spread spectrum for cell phones?
4. Explain the differences between cdma2000 and WCDMA? Is cdma2000 3G?
5. Name the two primary types of spread spectrum.
6. What is TD-SCDMA? Where is it used?
7. What organization works with the ITU to develop spread spectrum standards for cell phones?
8. What high-speed techniques are used with WCDMA to provide faster data connections in 3G cell phones?
9. What popular wireless technology uses frequency-hopping spread spectrum?
10. Define coding gain that occurs in spread spectrum systems.

Project 11-5 Orthogonal Frequency Division Multiplexing

Objective

To further explore the technology of OFDM and its current usage.

Required

PC with Internet access.

Introduction

Orthogonal frequency division multiplexing (OFDM) is one of the fastest-growing communication technologies. It appears that it could be the single most important method in future wireless applications. The purpose of this project is to give you the chance to get a more comprehensive view of OFDM and its current usage.

Procedure

1. Go to the following websites and explore the information on OFDM. Take any interesting tutorials and print out other information such as white papers.

   www.wikipedia.org
   www.sss-mag.com
   www.palowireless.com
   www.commsdesign.com

2. Perform Internet searches on OFDM and related topics and examine the various documents produced.
3. Answer the questions below, using the resources you discovered.

Questions

1. Explain in detail just what "orthogonal" means with respect to OFDM.
2. In wireless applications, what characteristic of OFDM improves reliability and performance?
3. Name three current wireless technologies or standards that use OFDM.
4. What is the name of the OFDM-like technology used in wired broadband systems?
5. What mathematical technique is at the heart of all OFDM?

Project 11-6 Digital Subscriber Line

Objective

Expand your knowledge of digital subscriber line and its application.

Introduction

Digital subscriber line (DSL) is the most widely used broadband technology in the world. It is widely used for Internet access over the standard telephone lines. Over the years, it has grown in usage and new standards have been developed. Because of its widespread usage, it is important to understand this important technology. This project lets your explore the details of DSL and its variants.

Procedure

1. Go to the following websites and explore the information on DSL. Take any interesting tutorials and print out other information such as white papers.

   www.wikipedia.org
   www.sss-mag.com
   www.palowireless.com
   www.commsdesign.com
   www.dslforum.org

2. Perform Internet searches on DSL and related topics and examine the various documents produced.
3. Answer the questions below, using the resources you discovered.

Questions

1. What is the name of the modulation technique used in DSL? What is the equivalent wireless technology?
2. What international organization develops standards for DSL?
3. Make a table of the main variations of DSL, giving the standards designation, upper data rate, and range of transmission.
4. What part of the local loop spectrum is used by ADSL, ADSL2+, VDSL2?
5. List the primary limitations, faults, and variations of the local loop that restrict higher data rates and distances on the local loop.

Project 11-7 OSI Model

Objective
To become more familiar with the Open Systems Interconnect model.

Introduction
The Open Systems Interconnect model is a design that organizes all of the functions of a network from the communications link to the application. It is based on a layered concept with each of its defined seven layers designated to handle specific parts of the networking operations. Most modem networking systems and standards follow the OSI model or a variation thereof. It is helpful to understand this model in learning and working with networks. This project provides a way to expand your knowledge of the OSI model beyond the introduction in the text.

Procedure
1. Go to the following websites and access the tutorials in each.

   www.inetdaemon.com/tutorials/theory/osi/index.shtml
   www.cisco.com/univercd/cc/td/doc/cisintwklitodoc/introint.htm
   www.PcsuPPortadvisor.com/OSI7Iayermodel/pagel.htm
   computer.howstuffworks.com/osi.htm

2. Do an Internet search on the OSI model. Print out and read relevant material.
3. Print out each tutorial.
4. Read the tutorials on the OSI model.
5. Answer the Review Questions.

Questions
1. Name the lower layers of the OSI model.
2. Name the upper layers of the OSI model.
3. Designate which layers are hardware and which are usually software.
4. Are all layers used in each network? Explain.

Chapter

12

Networking, Local Area Networks, and Ethernet

Project 12-1 Exploring Ethernet

Objective

To expand your knowledge of Ethernet, its operation, and its applications.

Required

PC with Internet access.

Introduction

The most widely used LAN technology is Ethernet. And Ethernet has expanded beyond LANs to metro networks and wireless. Ethernet is so ubiquitous that a working knowledge of it is essential to any work that you may do in communications. This project will let you dig deeper into the types of Ethernet, its operation, and its usage.

Procedure

1. For a good general review of Ethernet, go to the following:
   a. www.wikipedia.org/wiki/IEEE_802.3
   b. www.wikipedia.org/wiki/Ethernet
2. Go to the following websites and explore the information available on Ethernet.

   grouper.ieee.org/groups/802/3
   www.ieee802.org
   www.ethernetalliance.org
   www.metroethernetforum.org
   www.iol.unh.edu/services/testing/ge/training/
   www.lantronix.com/learning/networking.html

3. Perform a Web search on the terms Ethernet, Ethernet tutorial, and any related terms you encounter in the questions. Examine the material you find, take any relevant tutorials, and print out information of interest.
4. Answer the questions below about Ethernet.

Questions

1. What organization standardizes Ethernet? What is the standard designation?
2. What is the standard designation (802.?) for Gigabit Ethernet?
3. What does the term MAC mean with regard to Ethernet. Explain the MAC.
4. What are the five main data speeds of Ethernet? What is the primary medium?
5. Can data be transferred over twisted pair at 10 Gbps? Explain.

6. Name the fiber optical versions of Ethernet.
7. What does XAUI mean? Where is it used?
8. What is metro Ethernet?
9. Explain how Ethernet is used in iSCSI and LXI.
10. What is Ethernet in the first mile? What is EPON?
11. List the versions of 100 Gbps Ethernet.
12. State the purpose of the following forthcoming Ethernet standards: 802.3an, 802.3aq, 802.3ap, 802.3at, and 802.3as.
13. Can Ethernet be used as a MAN? WAN?

Project 12-2 Software Networking

Objective

To examine some new approaches to networking.

Required

PC with Internet access.

Introduction

Networking is a combination of both hardware and software. However, as network speeds and size increase, many believe that network performance can best be improved with software methods. Two of these approaches are software-defined networking (SDN) and network functions virtualization (NFV). This project lets you become familiar with these new concepts.

Procedure

1. Search on the term software-defined networking.
2. Search on the term network functions virtualization.
3. Answer the questions below.

Questions

1. Define SDN.
2. What is virtualization?
3. Define NFV.
4. What are the advantages of SDN and NFV?
5. How are the two technologies similar or different?
6. How will these new approaches affect Ethernet?

Chapter 13

Transmission Lines

Project 13-1 Measuring SWR

Objective

To use a standing wave ratio (SWR) meter to measure the SWR of a coax transmission line.

Required

1   50-ft length of coax cable (RG-8x or RG-58A/U) with PL-259 UHF connectors
1   SO-239 UHF coax connector
1   51-Ω resistor
1   33-Ω resistor
1   100-Ω resistor
1   MFJ-269 SWR analyzer
1   N-to-UHF coax adapter
1   solder lug

Introduction

This project will allow you to use a commercial meter for measuring SWR on a coax cable. That is a common test on antennas. The recommended instrument is the MFJ-269 analyzer, which will measure SWR over a frequency range of 1.8 to 470 MHz. The MFJ-269 has a built-in signal generator for this frequency range, a frequency counter, and the SWR and power measuring circuitry.

To prepare for this project, read the instruction manual for the MFJ-269 to learn how it operates.

Procedure

1. Using the SO-239 coax connector, build a load for the coax cable. Mount a solder lug on one corner of the connector with a #4 screw and nut as shown in Fig. 13-1. Solder the 51-Ω resistor to the center conductor, of the connector and the solder lug as shown. Screw the load onto one of the PL-259 UHF connectors on one end of the coax cable.
2. Given the cable and generator impedances and the load impedance, what load impedance and SWR do you expect to measure?
3. Attach the N-to-UHF adapter to the MFJ-269 output connector, then screw the UHF connector on the free end of the coax cable to this adapter.
4. Apply power to the MFJ-269. Rotate the band switch to the 10–27-MHz range. Then set the frequency to 12 MHz.
5. Read the load impedance and SWR from the MFJ-269 display. How do these figures compare with what you predicted? Explain any variations.

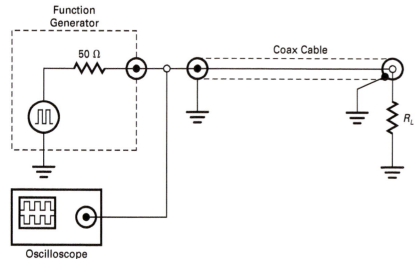

Fig. 13-5 Test setup for TDR. Note that the oscilloscope monitors the input signal.

so no reflection occurs. The signal at the input is not affected, because there is no reflection. In these illustrations, we are assuming no signal attenuation in the cable.

In Fig. 13-6(b), the line is open at the end, thereby producing 100 percent reflection. The reflected pulse will appear back at the input delayed by the line where it adds to the input pulse. Note the time shift $(2t_d)$ between the generator input signal and the reflected signal. Keep in mind that reflected pulses experience the same delay in traveling from the load back to the input. The reflected signal Y adds to the generator signal X algebraically, producing a composite signal at the cable input.

In Fig. 13-6(c), the line is shorted at the load. This also results in 100 percent reflection, but the returned pulse is inverted in polarity as well as being shifted in time. Note that the resulting signal at the input is bipolar when the reflected signal is added to the generator signal.

Figure 13-6(d) shows the waveform resulting when the load impedance (Z_L) is greater than the line impedance and generator impedance (Z_0), or $Z_L > Z_0$. Figure 13-6(e) shows the waveform resulting when the load impedance (Z_L) is less than the line impedance and generator impedance (Z_0), or $Z_L < Z_0$. These waveforms are only approximations, as the exact shape of the composite input signal depends on the length of the line.

In this experiment, you will demonstrate these principles and observe the waveforms resulting from TDR.

Procedure

1. Use a piece of RG-58/U coax cable whose length is unknown. The length should be greater than 40 ft but does not have to be longer than about 100 ft. Strip both ends of the cable, exposing the center conductor and the shield braid. Connect small alligator clips to one end of the line to make it quick and easy to connect different values of load resistance. Connect the other end of the cable to the function generator output. You may wish to use a BNC connector on this end of the cable, as most function generators have a BNC output connector. Use a BNC T-connector to provide an output to the oscilloscope. See Fig. 13-7.

2. The function generator should be set to supply 1-MHz square waves. The output should supply dc pulses, and the output impedance of the generator

134

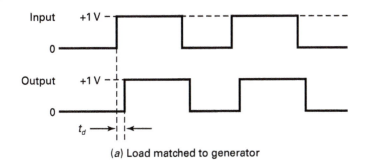

(a) Load matched to generator

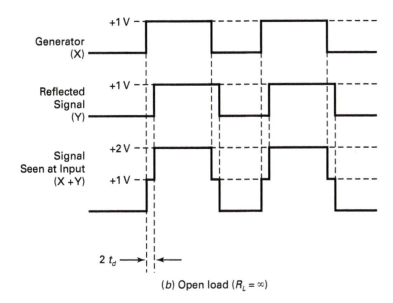

$2\,t_d$ ⟶

(b) Open load ($R_L = \infty$)

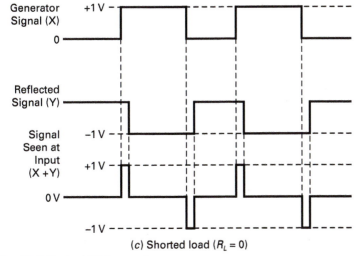

(c) Shorted load ($R_L = 0$)

Fig. 13-6 Typical TDR waveforms.

should be 50 Ω. Most function generators have a TTL output that will be ideal for this experiment.

3. Display the function generator output signal on the oscilloscope and measure the generator voltage. it should be somewhere between 3 and 4 $V_{p\text{-}p}$ if it is TTL compatible. Record your value below.

$V_{in} =$ _____

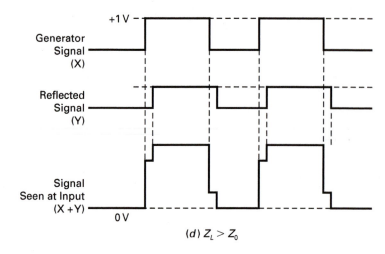

(d) $Z_L > Z_0$

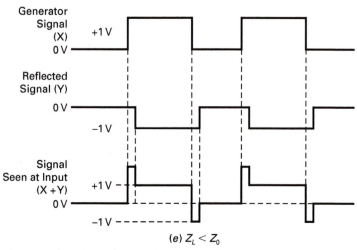

(e) $Z_L < Z_0$

Fig. 13-6 (continued) Typical TDR waveforms.

4. Now, connect the circuit shown in Fig. 13-5. Use a 51-Ω resistor for the load. Observe the input signal. Measure the voltage value on the scope and record your value below. Explain what you see. Is the load matched to the line?

V_{in} = _____

5. Using the second scope input channel, display the signal across the load along with the input signal on the other channel. Measure the output voltage and record below. Is this voltage value correct for a matched line?

V_o = _____

6. While observing the input and output signals, adjust the trace positions to overlap the signals. Measure the time shift between the two signals.

t_d = _____ ns

7. Using the formulas and values given earlier, calculate the equivalent inductance per foot for this cable. Then use the L and C values to calculate the time delay of 1 ft of the line.

L = _____ μH/ft

t_d = _____ ns/ft

8. Measure the coax cable length with a tape measure. Then multiply it by the time delay per foot you calculated in step 7 to determine the overall cable delay.

Cable length = _____ ft

t_d = _____ ns (total)

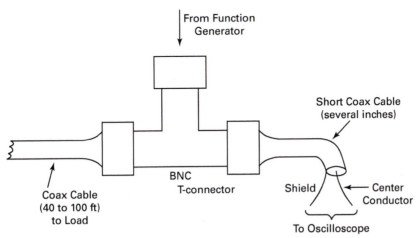

Fig. 13-7 Using a BNC T-connector to make connections between generator, oscilloscope, and cable.

9. Compare your measured and calculated values and account for any differences.

10. Remove the 51-Ω load. With the line open, observe the signal at the line input. Draw what you see. How does it compare to the waveform given in the Introduction?

11. Next, short the end of the coax. Again, observe the line input. Draw the waveform. How does it compare to the illustrations given in the Introduction?

12. Unshort the line and connect a 100-Ω resistor. Observe and draw the input waveform. Explain. What is the SWR for this arrangement?
    SWR = _____

13. Remove the 100-Ω load and connect a 22-Ω load in its place. Once again, look at the input waveform and draw it. Explain. Calculate the SWR.
    SWR = _____

Questions

1. Looking at the waveforms you drew in steps 10, 11 and 12 (which should be similar to these in Fig. 13-6), explain how you could calculate the length of the line from measurements you could make from these waveforms.

2. What is the SWR of an open or shorted transmission line?

3. A 75-ft piece of RG-58/U coax is properly terminated. A sine wave with a frequency of 2 MHz is applied to the line. What is the phase shift between the input and output?

4. A buried coax cable is used to carry cable TV signals from the cable head end to each subscriber. The coax cable between the subscriber's house and the nearest cable amplifier box is accidentally cut when a utility company is digging nearby. Explain a way that the cable company technician could use TDR to estimate the distance to the cable break.

Project 13-5 Smith Chart Tutorial

Objective

To review the use of the Smith chart.

Introduction

Despite its age (invented in 1938), the Smith chart is still widely used in electronics, especially in the design and application of transmission lines. It is a complex chart that takes some practice to master. The text covers this chart in some detail, and this project gives you the opportunity to review that material and to discover what other materials are available.

Required

PC with Internet access.

Procedure

1. Go to the following websites and explore the materials available. Print out and read any that are interesting. Look for material on how Smith charts are used in impedance matching.

   www.sss-mag.com
   en.wikipedia.org/wiki/Smith_chart

2. Do a Google, Yahoo, or Bing search on the Smith chart. Also search on Smith chart tutorial. Find and read any tutorials that you find interesting. Again, look for material on how Smith charts are used for impedance matching.

3. Write a short paper on how Smith charts can be used to match one impedance to another. Give one worked-out example.

Chapter

14

Antennas and Wave Propagation

Project 14-1 Dipole Radio Antenna

Objective

To construct a dipole antenna and test its directional characteristics.

Required

1   FM radio with terminal for attaching one of the following external
    antennas:
- 300-Ω twin lead (about 8 to 15 ft), or
- #14 solid copper wire with RG-59/U 75-Ω coax cable and matching
  RCA phono connector
- Soldering iron, solder, side cutters, and knife or wire stripper

Procedure

1. Determine what type of antenna terminals are used on the FM receiver.
   If a 300-Ω twin lead is used, two screw terminals will be available. If
   coax is used, the connector is most likely the RCA phono type.
2. Use the twin lead to construct a folded dipole antenna for your favorite
   FM radio station, whose frequency f you must know. What is it?
3. Cut the antenna following the guidelines in the text. It must be one-half
   wavelength long, but take into account the velocity factor of twin lead,
   which is 0.8. What length do you need?
4. Strip both ends of the twin lead and twist the leads together, as Fig. 14-1
   shows. Cut into the center of one conductor of the twin lead and attach
   the 300-Ω transmission line; see Fig. 14-1. The transmission line length
   is not critical, but it should be great enough to attach the line to the receiver

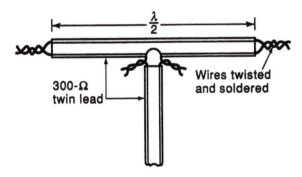

Fig. 14-1 Folded dipole construction with twin lead.

Project 15-2 Review of SONET

Objective

To refresh your knowledge of SONET.

Introduction

SONET, as you know, refers to the Synchronous Optical Network, a fiber-optic system that is the heart of the Internet backbone and most telecommunications long-distance operations. Because it is so widely used, it is essential for you to have a good basic knowledge of it. The text gives an overview, but this project allows you to explore SONET in more detail.

Required

PC with Internet access.

Procedure

1. Perform Internet searches on SONET and related topics and examine the various documents produced. Look at the Wikipedia entry at www.wikipedia.com.
2. Do any other searches you need to answer the questions below.

Questions

1. Can Ethernet packets be transmitted over SONET? Explain.
2. What is SDH? How is it different and similar to SONET?
3. What is the current maximum data rate of SONET? What is the designation?
4. What are ODAM and RODAM?

Project 15-3 Review of OTN

Objective

To expand your knowledge of OTN.

Introduction

OTN is the optical transport network. This new standard and protocol is replacing SONET and other optical technologies in the Internet. It offers some major benefits over SONET and will eventually be the primary backbone of the Internet. This project lets you dig deeper into OTN.

Required

PC with Internet access.

Procedure

1. Search on the terms OTN and optical transport network. Look for tutorials and short summaries.
2. Answer the questions below.

Questions

1. What standards organization develops and maintains OTN?
2. What are the primary data rates of OTN? The maximum?
3. What FEC is used in OTN?
4. State two main benefits of OTN over SONET.
5. What networking product or equipment terminates OTN?

Chapter 17

Satellite Communication

Project 17-1 Exploring GPS

Objective

Review the operation and application of GPS and its alternatives.

Introduction

The Global Positioning System (GPS) is the premier satellite navigation system in the world. Not only is it used for military navigation, but also it has become a major commercial and personal benefit as new applications have been found and low-cost GPS radios have become available. Today, GPS is so widely used that it plays an active role in our everyday lives. This project provides a way to learn more about it.

Required

PC with Internet access.

Procedure

1. Perform an Internet search on GPS tutorial. Print out and read any material you find. Don't forget to look at Wikipedia at www.wikipedia.com.
2. Go to the websites of the major GPS receiver manufacturers such as Garmin, Trimble, Magellan, TomTom, and others and explore the material available such as white papers and application notes. Also look at some of the specific products available and their applications.
3. Answer the questions below.

Questions

1. Name two other current and future global navigation satellite systems (GNSS).
2. List 10 specific applications for GPS.
3. Is GPS used in cell phones? Explain.
4. Many GPS users fear that by having a GPS receiver, others can track them. Is that true? Explain.
5. What is geocaching?
6. What is assisted GPS?

Chapter
19

Optical Communications

Project 19-1 Optical Fiber Data Transmission

Objective

To demonstrate the transmission of binary data over a simple fiber-optic link.

Required

1   oscilloscope
1   function generator (TTL square wave)
1   fiber-optic cable (3 ft), 1000 micrometer (1 mm) diameter
1   IR LED, with housing and connector (Industrial Fiber Optics IF-E91A-R)
1   IR phototransistor, with housing and connector (Industrial Fiber Optics IF-D92)
2   NPN transistors (2N3904, 2N4401, or similar)
1   100-Ω resistor
1   1-kΩ resistor
1   3.9-kΩ resistor
1   33-kΩ resistor
1   220-kΩ resistor

Introduction

More and more data communication is being carried out over optical fiber cables rather than via twisted-pair, coax, or other wire conductor cables. This includes large communications systems such as the telephone network as well as smaller systems such as local area networks. High speed, high capacity, wide bandwidth, low noise, high security, and other characteristics give fiber-optic equipment major advantages over older and more conventional data communications techniques. If you plan to work in the electronics communications field, you will most surely encounter fiber-optic equipment. The purpose of this experiment is to familiarize you with the basic elements of all fiber-optic systems.

First you will examine the fiber-optic cable itself and note its construction and characteristics. You will learn how to cut the end of a cable and mount it in connectors. Next you will construct an LED transmitter and send light pulses down the cable. Finally, you will build a receiver with a photo transistor and demonstrate the reception of the light pulses. This experiment will give you a good feel for the hardware and operation of typical fiber-optic equipment.

Procedure

1. Locate the fiber-optic cable. Use a wire stripper to remove about ⅛ in. of outer insulation to expose the fiber element itself. Note the clear plastic fiber. Look at it under a magnifying glass if one is available.

2. Point one end of the cable at a strong light source—a window (daylight, of course), a lamp, an overhead light fixture, a flashlight, or whatever is convenient. Observe the other end of the cable, and you will see the light being transmitted. Put your finger over the end of the cable pointed at the light and note how the light at the other end disappears. Point the end of the cable at several light sources with different intensities and see how the light at the other end varies. Experiment further as desired.

3. Construct the transmitter circuit shown in Fig. 19-1. The physical connections to the transistor and LED also are given. Do *not* connect the cable to the LED yet.

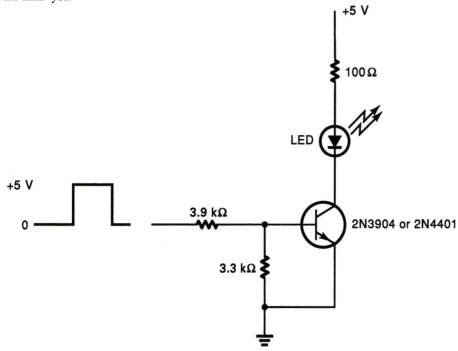

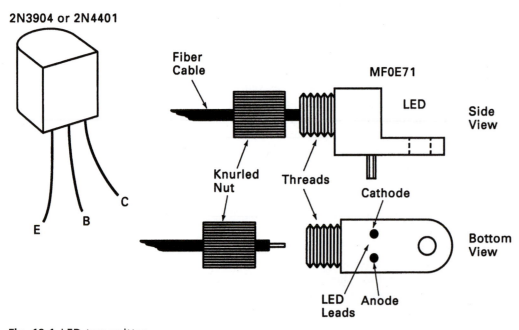

Fig. 19-1 LED transmitter.

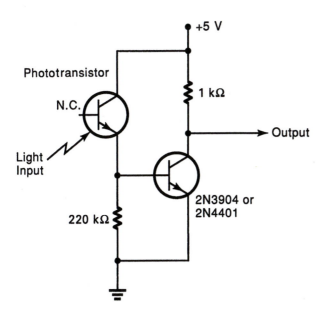

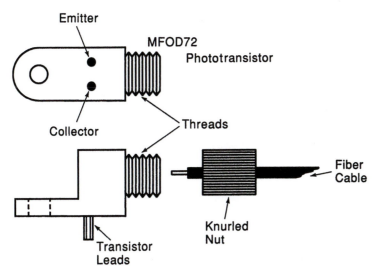

Fig. 19-2 Receiver circuit.

4. Connect a TTL square wave from a function generator to the input of the transmitter and set the transmitter frequency to about 1 Hz. Apply power to the transmitter circuit and look inside the hole where the fiber cable goes. You should see the faint red glow of the LED switching off and on at a 1-Hz rate.

5. Turn off the power and wire the receiver circuit as in Fig. 19-2.

6. Prepare the fiber-optic cable by stripping about ⅛ in. of outer sheath from each end. Use a sharp razor blade or X-Acto knife to cut the end of the fiber clean and square. Rotate the cable as you cut so that you get a flat, even end surface.

7. Remove the knurled nut from the LED housing and slip it over one end of the fiber-optic cable. Then insert the fiber into the LED housing as Fig. 19-1 shows. Push the fiber cable all the way in and then tighten the nut.

8. Repeat the above procedure for the other end of the cable and install the cable in the photo transistor assembly.

9. Set the function generator to supply a square wave with a frequency of 1 kHz. Assuming that input, predict the output from the receiver by studying the transmitter and receiver circuits.

10. Apply power to the circuit. Connect an oscilloscope to the receiver output. What signal do you see? Does it conform to your prediction?

11. What is the receiver output when the transmitter input is 0 (gnd)? What is it when the receiver input is 1 (+5 V)? Record both results.

12. While observing the receiver output, begin increasing the function generator frequency slowly beyond 1 kHz. Describe what happens. Explain why. What is the maximum frequency before the output begins to decrease in amplitude?

13. Verify that it is the light over the fiber-optic cable that is operating the receiver. Disconnect the cable from the transmitter end while observing the pulse output at the receiver. Describe what you see. How close does the cable have to be to the LED to see an output?

14. Can analog signals be transmitted over a fiber-optic link?

Project 19-2 Infrared Remote Control

Objective

To demonstrate infrared (IR) remote control.

Required

1 oscilloscope
2 power supplies, +12 V and −12 V
1 remote control unit from a TV set, VCR, etc.
1 IR diode (940-nm wavelength), IRD500 or equivalent
1 op amp, general-purpose (741, LF411, AD548, etc.)
2 0.1 μF ceramic disk capacitors
1 10-MΩ
• Batteries for the remote control unit (usually 2 AAA type)

Introduction

IR (infrared) remote control units are used on virtually all TV sets today. They are also used in many other applications, including remote controls for radios, stereo systems, CD players, ceiling fans, VCRs, and cable boxes. IR is also used in computer systems for wireless local area networks (LANs) and as a wireless interface to printers and personal digital assistants (PDAs). IR is cheap and easy to use.

The digital data to be serially transmitted is applied to a modulator that transmits the binary 1s and 0s as bursts of high-frequency pulses. These pulses then turn an infrared LED off and on. Different types of binary codes have been developed to represent numbers and control functions (volume, channel change, etc.). As the button on a remote is pressed, the transmitter sends the binary code for that digit or function again and again.

In this experiment, you will build an IR receiver and observe the digital codes produced by a typical TV remote control.

Procedure

1. Construct the circuit shown in Fig. 19-3. The IR sensitive diode operates with reverse bias. The IR diode looks like a clear LED. The short lead is

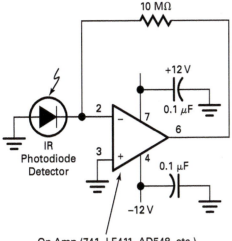

Fig. 19-3 IR receiver.

the cathode, and the cathode is the lead nearest the flat side of the housing. When infrared light strikes the diode, it conducts and a small current flows. This current is amplified by the op amp and produces an output. You will monitor the op amp output on the oscilloscope.

2. Apply power to the circuit. Connect the scope to the op amp output.

3. Point the remote control at the IR diode and press any key. Be sure that the remote is less than 1 ft away from the diode. Hold the key down continuously while you adjust the oscilloscope for a stable display. Experiment with the position of the remote with respect to the diode. Adjust the position for maximum output voltage on the scope. What do you see?

4. While continuously pressing one of the buttons on the remote control unit, adjust the scope sweep to display the serial binary code. You will have to play around with the sweep setting and triggering control to get a stable display.

5. Measure the pulse spacing and pulse width of the pulses you observe.

6. Explain how the pulses differ with different buttons depressed.

   Note: To observe the signal on the scope, you will need to keep one of the buttons on the remote unit pressed continuously. This will considerably drain the batteries in the remote. If the signal gets weak or the unit "dies," you can assume that the batteries must be replaced.

7. Press the 1 through 9 buttons and try to discern the binary code.

8. While pointing the remote at the receiver, back away and increase the distance between you with the remote and the receiver. Continue increasing the distance until the signal is no longer shown on the scope. You may need a partner to observe the scope and tell you when the signal disappears. Measure the maximum distance before the signal drops out.

9. While pointing the remote directly at the IR diode with about 5 ft separating the two, begin moving horizontally to the left in a circular arc while still keeping the diode in the same plane. See Fig. 19-4. At some point, the signal will disappear. At this point, estimate the angle between the centerline and the line where the signal dropped out. Repeat this procedure to the right. What is the approximate total view angle of the IR diode?

Discussion Topics

1. What do you think are the three most commonly pressed buttons on a remote control unit?

2. Explain the results you got in step 8 of the Procedure.

3. If another remote control from a different product is available, investigate its performance by repeating steps 2 through 6. Describe your results. Are the pulses and codes the same?

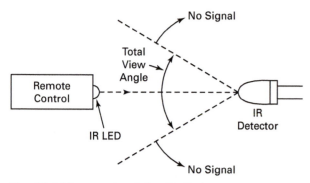

Fig. 19-4 View angle of the IR diode.

Project 19-3 Fiber Optic Tutorial

Objective

To further explore fiber-optic communications systems and dense wavelength division multiplexing.

Introduction

The fastest communications and networking techniques use fiber-optic methods. High-speed lasers transmit the data over glass or plastic fiber cable at speeds to 100 Gbps and higher. Further bandwidth increases are coming from a frequency division multiplexing method called dense wavelength division multiplexing (DWDM). It uses different wavelength, frequencies, or "colors," of light to transmit multiple high-speed streams on the same fiber. This project gives you a review of the basic fiber-optic communications methods and further details on DWDM.

Required

PC with Internet access.

Procedure

1. Perform a search on the terms fiber optic technology, dense wavelength division multiplexing (DWDM), polarization mode and chromatic dispersion, and similar terms. Seek out tutorial material.
2. Answer the questions below.

Questions

1. Why doesn't fiber replace twisted pair in the local loop?
2. What is the term describing the connection for the consumer's home by fiber-optic cable?
3. What is the name of the short-range fiber network that is beginning to provide fiber to consumers? Name two standards associated with them.
4. What are the primary factors that restrict the data rate and distance on fiber-optic cable? How is it overcome?
5. Name two standards used for 100 Gbps fiber systems.

Chapter

20

Cell Phone Technologies

Project 20-1 Contemporary Wireless, Part 1

Objective

Review the most commonly used cell phone technologies.

Introduction

The largest part of the wireless industry today is the cell phone business. Cell phones and the infrastructure that supports it use just about every communications and networking technique ever invented. This project provides a general introduction to the processes and techniques used to provide cell phone service worldwide.

Required

PC with Internet access.

Procedure

1. Go to the website www.work-readyelectronics.org.
2. At the top of the page, click on the Modules box, which will take you to a second page listing all the various modules you can access for free.
3. Answer the three questions given, then select the Contemporary Wireless module.
4. When you have accessed the module, follow the instructions given and complete the first three sections of the module: Cellular Telephone Systems, Access and Duplexing, and Cell Phone Technologies. The remaining two sections will be covered in a later project.
5. Be sure to take advantage of the Knowledge Probes that help you to review the material.
6. Take the formal Assessment at the end of the module.

Project 20-4 Small Cell Networks

Objective

To investigate the small cells movement for future cellular expansion.

Introduction

Small cells are miniature cellular base stations that are designed to expand existing cellular telephone systems. They are referred to by the term *heterogeneous networks* or HetNets. They offer several important benefits over the larger standard macro base stations. This project will help you understand the basics of small cells.

Required

PC with Internet access.

Procedure

1. Search on the terms small cells, heterogeneous networks, or similar terms. Seek tutorial materials.
2. Answer the questions below.

Questions

1. List the four major sizes or types of small cells and their basic caller capacities.
2. What is the main rationale for small cells? What problems to they solve, and what are the benefits?
3. Do small cells replace macro base stations?
4. What is the radio technology of small cells?
5. How are small cells powered?
6. What is the backhaul for small cells?
7. Is a Wi-Fi access point a small cell?
8. What is SON?

Chapter 21

Wireless Technologies

Project 21-1 Contemporary Wireless, Part 2

Objective

Review popular wireless methods and services.

Introduction

While cell phones dominate the wireless business, there are still many other contemporary wireless technologies and services. Most of these were covered in Chapter 21 of the text. This project provides you with a review of the most commonly used of these short-range wireless methods.

Required

PC with Internet access.

Procedure

1. Go to the website www.work-readyelectronics.org.
2. At the top of the page, click on the Modules box, which will take you to a second page listing all the various modules you can access for free.
3. Answer the three questions given, then select the Contemporary Wireless module.
4. When you have accessed the module, follow the instructions given and complete the remaining two sections of this module: Wireless Local Area Networks and Short-Range Wireless Technologies.
5. Be sure to take advantage of the Knowledge Probes that help you to review the material.
6. Take the formal Assessment at the end of the module.

Project 21-2 Comparing Short-Range Wireless Technologies

Objective

To examine the most widely used short-range wireless technologies in more detail.

Introduction

There are literally dozens of different wireless technologies in use today. The most popular are Wi-Fi, Bluetooth, ZigBee, NFC, RFID, and ISM band. Each seems to have targeted a specific application niche. And each has unique characteristics. In this project, the object is to dig deeper to get a more thorough understanding of the various technologies available.

Required

PC with Internet access.

Procedure

1. Using Internet searches and any other available resources, make a table listing the six technologies listed in the introduction and give the following information for each:
   a. Operating frequency range
   b. Maximum data rate
   c. Standard designation and standard organization, if applicable
   d. Modulation method(s)
   e. Maximum range
2. For each of the six wireless technologies you identified above, list its primary application.
3. Search for any new wireless standard not listed in Chapter 21 of the text or in this project. List at least two.
4. What FCC regulation generally applies to these technologies?
5. What short-range wireless technologies are usually found in smartphones?

Project 21-3 Wireless Monitoring

Objective

To demonstrate how physical conditions can be monitored wirelessly.

Introduction

There are many applications that require only simple monitoring of physical conditions. Some examples are the monitoring of open windows or doors or the position of an object. A single switch can be used as a sensor. A roller lever switch could be used to sense an open garage door. A magnetic reed switch would make a good sensor on a window. Special limit switches are available for a wide range of indications. If the switch is open, it signals that the door or window is closed. However, if the switch closes, this is an indication that the door or window is open. The closed switch can then send a signal to indicate this fact.

The wireless transmitter and receiver modules used in Project 14-2 can be used to make a wireless monitoring system. Figure 21-1 shows the transmitter diagram, and Fig. 21-2 shows the receiver. If the transmitter switch is open, no power is applied to the transmitter circuits, so no signal is sent. Closing the switch applies power to the circuit. A 555 timer IC supplies a 1.5 Hz rectangular wave to the TX3A transmitter. This is indicated by a flashing LED. The transmitter is frequency modulated by the 555 rectangular pulses and sends a 914 MHz signal to the receiver. The receiver picks up the signal, demodulates it, and supplies the recovered rectangular wave to a transistor that flashes an LED indicating an open condition.

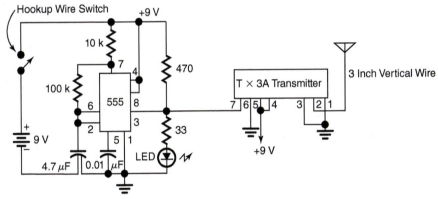

Fig. 21-1 Transmitter circuit.

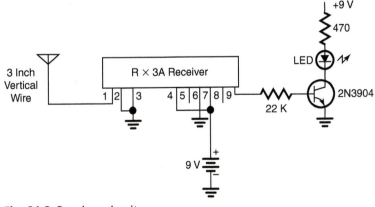

Fig. 21-2 Receiver circuit.

Required

1 Radiometrix TX3A-914-64 UHF transmitter
1 Radiometrix RX3A-914-64 UHF receiver
2 breadboarding sockets, one each for transmitter and receiver
2 power supplies for transmitter and receiver (9V), battery with battery terminal clips
1 555 timer IC
2 LEDs
1 2N3904 NPN transistor or equivalent
1 100-kΩ resistor
1 10-kΩ resistor
2 470Ω resistor
1 33Ω resistor
1 4.7 μF electrolytic or tantalum capacitor
1 0.01 μF ceramic capacitor
Several feet of #22 hookup wire to use as the switch contacts and as the antennas

Procedure

1. Construct the transmitter circuit on one of the breadboarding sockets. The antenna is simply a 3-in. length of hookup wire plugged vertically into one of the breadboarding connections that connect to pin 2 of the TX3A. A pair of hookup wires will serve as the switch. Just touch the two bare wires together to close the switch and apply power.
2. Test the transmitter by connecting the hookup wires together. The LED should flash. If it does, disconnect power.
3. Construct the receiver on the other breadboarding socket. Again, use a 3-in. piece of vertical hookup wire for the antenna on pin 1 of the receiver.
4. Test the receiver by applying power. Then turn on the transmitter nearby by closing the hookup wire switch. The receiver LED should flash. If all is OK, disconnect power.
5. A key test of the circuit is to separate the transmitter and receiver to see the maximum possible range.

Questions

1. What was the maximum range of the system? What limits it?
2. What FCC rules (which Part?) govern wireless on this frequency?

Chapter

22

Communication Tests and Measurements

Project 22-1 RF Test Equipment Familiarization

Objective

To explore the types of test equipment used in RF testing.

Introduction

Testing RF equipment is one of the most challenging parts of working with wireless systems. The reasons for this stem largely from the very high frequencies involved as well as from the complexity of most of the widely used systems. Furthermore, the equipment must conform to the rules and regulations of the FCC and of one or more standards organizations. The test instruments used for these tests and measurements are among the most sophisticated and expensive available. The purpose of the project is to become familiar with the key instruments used in such work.

Required

PC with Internet access.

Procedure

1. Using Internet access, go to the websites below.

   www.aeroflex.com
   www.anritsu.com
   www.keysight.com
   www.lecroy.com
   www.rhode-schwarz.com
   www.tektronix.com

   Examine the types of test instruments available for RF and wireless.
2. Using the information you gained from the sites listed above, and any necessary Internet search, answer the questions below.

Questions

1. Name the three primary test instruments used in RF test.
2. Which instrument, an oscilloscope or a spectrum analyzer, gives the most useful information about a complex modulated signal?
3. What arbitrary waveform generator (AWG) products have the highest output frequency? What is it? Can AWG signals include modulation?
4. What test instrument is used to produce an eye diagram?
5. What test instruments would you use to measure jitter? What is the unit of measurement for jitter?
6. What is a vector signal analyzer? What is it used to test?
7. What is the best instrument to be used to find EMI?

Appendix A

Breadboarding

CAUTION Breadboarding is generally limited to low-power circuits. High currents would cause high temperatures, and the breadboard would be damaged. High voltage could cause arcing and would be very dangerous. *If there is any doubt, consult your instructor before breadboarding a circuit.*

The name *breadboarding* has its origin in the early years of electronics, when circuits were sometimes fashioned on a flat piece of wood that looked like a breadboard (or actually was a breadboard). In case you didn't know, breadboards are used in some kitchens for working with dough or cutting bread. Today, wood breadboards are seldom used, and circuits are prototyped and tested using electronic breadboards, which are molded plastic with holes and internal metal contact strips. A circuit can be easily and quickly built by shoving component leads and jumper wires into the holes. Electronic breadboards are also called *prototyping boards*.

Figure A-1(*a*) shows how a series circuit appears on a typical electronic breadboard. Each group of five holes is electrically connected. Or, put another way, there is electrical continuity from any one contact to the other four contacts in that group. So, the top horizontal resistor is connected to the vertical resistor because each has a lead that is inserted into one group of five in Fig. A-1(*a*). Note that the vertical resistor and the bottom horizontal resistor are connected using another group of five.

There are also buses on breadboards. They are most often used for power, ground, or any signal requiring more than five common contact points. Figure A-1(*a*) shows two buses at the top and two at the bottom. Every contact along a bus is electrically connected to every other contact on that bus. Thus, the top source wire in Fig. A-1(*a*) is electrically connected to the top horizontal resistor via the top bus and jumper wire. Likewise, the bottom source wire is connected to the bottom horizontal resistor via the bottom bus and jumper wire. Note that the four buses in Fig. A-1(*a*) are electrically isolated from each other. So, one could use them as follows:

- +12-V bus
- −12-V bus
- +5-V bus
- ground bus

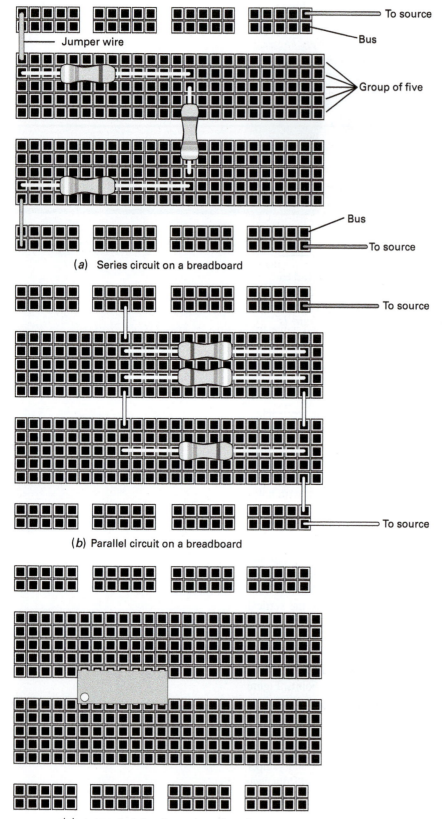

To source

Jumper wire

Bus

Group of five

Bus

To source

(*a*) Series circuit on a breadboard

To source

To source

(*b*) Parallel circuit on a breadboard

(*c*) Integrated circuit on a breadboard

Fig. A-1 Electronic breadboard.

Figure A-1(*b*) shows a parallel circuit. Notice how jumper wires are used to connect the top two resistors to the bottom resistor. There is a problem in Fig. A-1(*b*) in that both source leads are not connected to the resistors. Can you see why? How can the problem be corrected? Two possible answers are:

- Move the bottom source lead to the third bus down in Fig. A-1(*b*)
- Add a jumper wire from the third bus down to the bottom bus in Fig. A-1(*b*)

Figure A-1(*c*) shows that integrated circuits in a dual-inline package conform to the contact arrangement of the breadboard. The IC pins and the board contacts have the same center-to-center spacing of 0.1 in. Each IC pin in Fig. A-1(*c*) has four contact points available for interconnections. In cases when this is not enough, a jumper wire can connect an IC pin to an unused group of five, or two jumper wires to two groups, and so on.

Jumper wires are usually made from no. 22 or no. 24 solid, tinned, insulated wire. About $1/4$ in. of insulation is removed from each end. After a wire is used many times, the end might become bent and fragile. When this happens, cut off the damaged end. It is a good idea to use different wire colors, for example, red for positive, black for ground, green for negative, and white for signal paths. This becomes more of an issue when large, complicated circuits are breadboarded.

In classroom or lab settings, the component leads are usually left full-length, unless the ends are damaged as mentioned before. This is because the components are used over and over. This means that components might stand above the breadboard by an inch or more. This is usually acceptable, but be sure that they don't flop around and touch each other and cause short circuits.

Some component wires might be too big to fit comfortably into the contact openings. Do not force them. One solution is to solder a short length of no. 22 wire to the end of any wire too big in diameter for the breadboard. Some components cannot be soldered without ruining them. Fuses and lamps are examples. In these cases, fuse holders or lamp sockets will be required for soldering to the no. 22 wires. Consult your instructor in these cases.

Rules and Tips for Breadboarding

- High-power circuits should *not* be breadboarded.
- High-frequency circuits are usually not breadboarded. The higher the frequency, the more this is true.
- High-gain circuits might not perform well. There are often problems with noise and oscillation.
- Keep all component leads and wires as short as possible.
- Surface-mount components cannot be used without soldering leads to them or using special adapters (these are available for some surface-mount ICs).
- Arrange the placement of physical parts similar to the placement in the schematic drawing. This reduces errors and allows for easier checking. However, in the case of high-gain circuits, the output should be located away from the input regardless of how the schematic is drawn.
- Inventory all the parts before breadboarding. This reduces errors of omission and makes it less likely that an incorrect part will be used.
- Have your partner or a third party check your work for errors. He or she will often notice things that you will not.

- Work methodically. Wire all the power connections, then the grounds, and then the signal paths. Another method is to do resistors first, then capacitors, then ICs, and so on. Some people place a pencil check on the schematic for each interconnection and part as they are wired.
- Bypass the power bus or buses to ground with a 15- or 25-μF electrolytic or tantalum capacitor. Be sure to observe polarity and to use capacitors with an adequate voltage rating. Power-supply bypass capacitors will often correct strange behavior arising from noise or unwanted oscillation. The bypass capacitors should be located near any high-gain or high-speed logic devices that could cause a problem. For TTL and CMOS logic ICs, a 0.1-μF ceramic capacitor is recommended across the power to ground pins.

Appendix B
Lab Notebook Policies

The world is shrinking. More and more companies and organizations are becoming players in the global marketplace. These developments are changing the way companies do business. Questions of quality, reliability, and liability have made it necessary that all critical procedures be accurately documented and closely followed by all employees. Any procedure that can adversely affect the quality and/or the reliability of goods or services is a critical procedure.

Perhaps the most organized current worldwide effort to develop standards in this area is the set of guidelines known as ISO 9000, or ANSI Q9000. The abbreviation *ISO* stands for the International Organization for Standards, which is based in Geneva, Switzerland, and *ANSI* stands for the American National Standards Institute. The ISO and ANSI guidelines are one and the same. These guidelines are elaborate and will not be covered here. However, they are mentioned because they are one of the major driving forces behind many of today's work requirements. Many employers use policies and procedures that are a subset of national and international guidelines.

Technicians and engineers work together in many companies to experiment with new ideas, to test new designs, to simulate product performance, to verify safe performance under adverse conditions, to find and eliminate possible interference to other products, and so on. What is learned in the laboratory must be clearly documented. Most often, the findings are recorded in a lab notebook.

The lab notebook policies and procedures given here are generic. They do not match those of any given company. Probably, no single organization would require all of them. Your instructor might assign several lab experiments and ask that they be reported using some of the following procedures. Such an experience could be very helpful to you since more employers are adopting and using such procedures.

The lab notebook is provided by the company and shall be:

1. Permanently bound. Three-ring bindings and the like are not acceptable.
2. Paginated. Every page must be prenumbered, in ink, and pages must never be removed.
3. Titled. The cover or the first page must contain the company name, the location, the room number or lab number where it serves, the beginning and ending dates of use, and any special purpose that it might serve within a given lab (for example, an entire notebook could be devoted to safety testing only in a multipurpose lab).
4. Located in the lab. It is not permissible to take a lab notebook home. Photocopying of any type is also not allowed in most companies.

Lab notebook entries shall be:

1. In ink. Pencils are not to be used. Nothing may be erased, and no white-out or correction fluid may be used. Mistakes should be crossed out with a single stroke in such a way that they are still legible. All cross-outs must be initialed.
2. Clearly identified. Each job or procedure must have a title, a date, a job number, and the names of those engineers and technicians participating. Also, a customer name and/or a contract number might be required.
3. Complete. The information that is recorded should be brief but complete. The guideline often used is that the entire procedure could be duplicated by following only the information found in the notebook.
4. Traceable. Test equipment must be listed, including serial numbers and latest calibration dates. Material listings are sometimes extended to include complex parts such as ICs. The part history and all identification markings might be required. Since computers and computer software often play a major role in the lab, they are also listed and version numbers of the software are required.

In electronics labs, items often entered in the lab notebook include:

1. Statement of the purpose of the procedure.
2. Brief narrative describing the procedure.
3. Schematics and other relevant sketches.
4. Calculations and results. This includes documentation of the use of calculators and/or computers. Design calculations normally must be verified by independent means, and this too must be documented.
5. Results and/or data. This includes instrument readings, waveforms, and graphs. Use data tables where appropriate.
6. Notation of errors and possible explanations. Significant differences between calculated (theoretical) and measured values must be addressed.
7. References and related materials. This includes computer printouts, computer files, and equipment manuals. Small printouts might be taped or pasted into the lab manual. Or, the location and identification methods of the materials must be clearly stated. This policy varies from company to company. For example, in many organizations, some material will be found on a network file server.
8. Description of any equipment or component failures and possible explanations.
9. Any out-of-the-ordinary event that could possibly affect the results such as a lightning storm or power failure.
10. Conclusions and recommendations. If any part of the procedure was not entirely satisfactory, it needs to be stated here.

Don't forget to expect variations. Some companies require a table of contents in the lab notebook. This might be completed after the notebook is full. The table of contents can be added to pages 2 and 3, which have been reserved for this use.

Appendix C

GROL Review/Final Exam

As you learned in Project 1-1, the General Radiotelephone Operators License (GROL) is a credential issued by the FCC to those persons who want to work on selected marine and aircraft radio and other electronic equipment. To receive this license, you must pass an exam consisting of questions from Elements 1 and 3. Element 1 covers rules, regulations, and operating procedures. Element 3 covers electronic fundamentals, communications theory and circuits, and related equipment. The exam must be taken with one of several Commercial Operators License Examination Managers (COLEMs). You can find the details about the GROL and COLEMs at the FCC website:

wireless.fcc.gov/commoperators/index.htm?job=home

Look for the section Commercial Operator Home. You can review the types of licenses available and their requirements.

To help you prepare for the exam, the FCC provides a pool of sample questions for Elements 1 and 3. Element 3 also provides a good review of all the basic principles covered in the text related to this manual (*Principles of Electronic Communication Systems, 4th edition*) as well as a good summary of the related electronic fundamentals. Element 3 or parts thereof also make a good final examination for a course using this text.

The Element 3 question pool contains hundreds of questions divided up into sections. These are:

Subelement A Principles
Subelement B Electrical Math
Subelement C Components
Subelement D Circuits
Subelement E Digital Logic
Subelement F Receiver Theory
Subelement G Transmitters
Subelement H Modulation
Subelement I Power Sources
Subelement J Antennas
Subelement 3K Aircraft
Subelement 3L Installation, Maintenance, & Repair
Subelement 3MCommunications Technology
Subelement 3N Marine
Subelement 3O Radar
Subelement 3P Satellite
Subelement 3Q Safety

You should print out a copy of the Element 3 question pool. It is a Microsoft Word file. There is a group of questions for each of the subelements. Answers are given at the bottom of each page.

Subelements A through E cover fundamentals that are the prerequisite to a communications course. These are a good overview of the basics every communications engineer or technician should know.

Subelements F through J plus 3L and 3M cover material in the accompanying textbook.

The remaining subelements cover topics unique to the GROL license requirements.

Instructors

You can use these questions for chapter quizzes or on a final exam.

Students

You can use the questions as chapter reviews. You can also use the questions as a learning exercise. Give yourself an exam using the questions to see how you do. Review any topics you miss. Look up the answers to the questions you do not know in the accompanying text or another text, or do an Internet search.

Finally, plan to study for the FCC GROL exam. It is within reach with a little study. It is an excellent hiring credential as the license is widely recognized throughout industry.